UNDERSTANDING THE
LIMITS OF
ARTIFICIAL
INTELLIGENCE
FOR WARFIGHTERS

VOLUME 4_ WARGAMES

EDWARD GEIST

AARON B. FRANK

LANCE MENTHE

PREPARED FOR THE DEPARTMENT OF THE AIR FORCE
APPROVED FOR PUBLIC RELEASE; DISTRIBUTION IS UNLIMITED.

 PROJECT AIR FORCE

04

For more information on this publication, visit **www.rand.org/t/RRA1722-4**.

About RAND

RAND is a research organization that develops solutions to public policy challenges to help make communities throughout the world safer and more secure, healthier and more prosperous. RAND is nonprofit, nonpartisan, and committed to the public interest. To learn more about RAND, visit www.rand.org.

Research Integrity

Our mission to help improve policy and decisionmaking through research and analysis is enabled through our core values of quality and objectivity and our unwavering commitment to the highest level of integrity and ethical behavior. To help ensure our research and analysis are rigorous, objective, and nonpartisan, we subject our research publications to a robust and exacting quality-assurance process; avoid both the appearance and reality of financial and other conflicts of interest through staff training, project screening, and a policy of mandatory disclosure; and pursue transparency in our research engagements through our commitment to the open publication of our research findings and recommendations, disclosure of the source of funding of published research, and policies to ensure intellectual independence. For more information, visit www.rand.org/about/research-integrity.

RAND's publications do not necessarily reflect the opinions of its research clients and sponsors.

Published by the RAND Corporation, Santa Monica, Calif.
© 2024 RAND Corporation
RAND® is a registered trademark.

Library of Congress Cataloging-in-Publication Data is available for this publication.
ISBN: 978-1-9774-1281-2

Cover: Devrimb/Getty Images and Marc Andreu/Adobe Stock.

About This Report

This is the fourth report in a five-volume series addressing how artificial intelligence (AI) could be employed to assist warfighters in four distinct areas: cybersecurity, predictive maintenance, wargames, and mission planning. These areas were chosen to reflect the wide variety of potential uses and to highlight different kinds of limits to AI application. Each use case is presented in a separate volume, as it will be of interest to a different community.

This fourth volume describes the application of AI to wargaming and indicates which aspects of wargaming are most amenable to AI application and which are not. It is aimed at those with an interest in wargaming, the history of AI use in wargames, and the application of AI more generally. Volume 1 in the series provides a summary of the findings and recommendations from all use cases, and the other volumes provide detailed analysis of the individual use cases:

- Lance Menthe, Li Ang Zhang, Edward Geist, Joshua Steier, Aaron B. Frank, Eric Van Hegewald, Gary J. Briggs, Keller Scholl, Yusuf Ashpari, and Anthony Jacques, *Understanding the Limits of Artificial Intelligence for Warfighters:* Vol. 1, *Summary*, RR-A1722-1, 2024
- Joshua Steier, Erik Van Hegewald, Anthony Jacques, Gavin S. Hartnett, and Lance Menthe, *Understanding the Limits of Artificial Intelligence for Warfighters:* Vol. 2, *Distributional Shift in Cybersecurity Datasets*, RR-A1722-2, 2024
- Li Ang Zhang, Yusuf Ashpari, and Anthony Jacques, *Understanding the Limits of Artificial Intelligence for Warfighters:* Vol. 3, *Predictive Maintenance*, RR-A1722-3, 2024
- Keller Scholl, Gary J. Briggs, Li Ang Zhang, and John L. Salmon, *Understanding the Limits of Artificial Intelligence for Warfighters:* Vol. 5, *Mission Planning*, RR-A1722-5, 2024.

The research reported here was commissioned by Air Force Materiel Command, Strategic Plans, Programs, Requirements and Assessments (AFMC/A5/8/9) and conducted within the Force Modernization and Employment Program of RAND Project AIR FORCE as part of a fiscal year 2022 project, "Understanding the Bounds of Artificial Intelligence in Warfare Applications."

RAND Project AIR FORCE

RAND Project AIR FORCE (PAF), a division of the RAND Corporation, is the Department of the Air Force's (DAF's) federally funded research and development center for studies and analyses, supporting both the United States Air Force and the United States Space Force. PAF provides the DAF with independent analyses of policy alternatives affecting the development, employment, combat readiness, and support of current and future air, space, and cyber forces. Research is conducted in four programs: Strategy and Doctrine; Force Modernization and Employment; Resource Management; and Workforce, Development, and Health. The research reported here was prepared under contract FA7014-22-D-0001.

Additional information about PAF is available on our website:
www.rand.org/paf/

This report documents work originally shared with the DAF on September 23, 2022. The draft report, dated September 2022, was reviewed by formal peer reviewers and DAF subject-matter experts.

Acknowledgments

We thank our sponsor contact, Kathryn Sowers, and our action officers, Julia Phillips and Gregory Cazzell, for their guidance in choosing the use cases, for their thoughtfulness in scoping the research questions, and for working diligently with us to obtain the data necessary to conduct the many machine-learning experiments described in this series of reports. Thanks as well to R. Scott Erwin and Jean-Charles Ledé for graciously connecting us with many AI development efforts across the Air Force Research Laboratory.

We are also grateful to many current and former RAND colleagues, including Caolionn O'Connell, Sherrill Lingel, Osonde Osoba, and Chris Pernin for helping us shape the research agenda. Thanks to John Salmon for sharing his insights and to Matthew Walsh for reviewing this document. Great appreciation to Ellie Bartels for sharing her expertise on wargaming. We could not have written these reports without their help; any errors that remain are ours alone.

Summary

Issue

In the 2010s, rapid progress in artificial intelligence (AI) for game-playing inspired intense interest in the possible benefits of the technology for playing wargames. Advocates suggested that AI might make wargames more effective or make it possible to apply wargames to novel problems. The purpose of this study is to evaluate the probable limits of AI techniques for wargaming to identify more and less promising areas for future investment.

Approach

Extending earlier RAND Corporation research, this report specifies a taxonomy of wargames by type or purpose (*systems exploration, innovation, alternative conditions,* and *evaluation*) and by time-phased task (*preparing, playing, adjudicating,* and *interpreting*). We use these frameworks to assess the technical feasibility and cost-effectiveness of applying AI to various aspects of a given type of wargame under particular conditions.

Key Findings

Figure S.1 summarizes our assessment of the technical feasibility and cost-effectiveness of applying AI to wargames by type and task, following the extended taxonomy.

Figure S.1. Technical Feasibility and Cost-Effectiveness of Artificial Intelligence for Wargames

Type \ Task	Preparing	Playing	Adjudicating	Interpreting
Systems Exploration	Feasible if low-quality players are adequate for game application	Feasible for well-defined domains	Feasible when adjudication terms are straightforward	Requires human-level AI
Innovation		Requires human-level AI	Possibility of unforeseen player actions complicates adjudication	
Alternative Conditions		Feasible for well-defined domains	Feasible when adjudication terms are straightforward	Feasible except where differences are very complex
Evaluation		Feasible if low-quality players are adequate for game application	Feasible when adjudication terms are straightforward	Feasible except where goals are not well-specified

Prohibitive	Possible	Feasible

On the basis of this assessment and the considerations that underlie it, AI appears to be most promising for wargames that

- are designed to investigate *alternative conditions* or that are used for *evaluation*, especially wargames that address problems that are well specified and well bounded (i.e., those with a small set of well-defined evaluation criteria)
- employ computational models in a significant role during the adjudication process or in wargames that generate large volumes of digital data that must be adjudicated (e.g., cyber and electronic warfare)
- use advanced human-computer interaction (HCI) technologies for data capture and interaction (e.g., cameras and microphones) to create datasets for future AI applications
- are repeated many times, especially those that model zero-sum, force-on-force conflicts.

In contrast, AI appears to be least promising for wargames that

- are designed for *systems exploration* or *innovation*
- employ limited digital infrastructure or interaction with computational models
- are conducted in security environments in which advanced HCI technologies are restricted
- are played as one-offs or a very limited number of times for specific purposes.

Recommendations

1. **Resources that are devoted to developing AI applications for wargames should be concentrated on the most-promising areas.** This includes areas to investigate alternative conditions or that are used for evaluation, with well-defined problems and criteria; areas that

already incorporate digital infrastructure, including HCI technologies; and areas that are regularly repeated, such as force-on-force conflicts.

2. **The use of digital gaming infrastructure and HCI technologies should be increased,** especially in games designed for systems exploration and innovation. The digitization of wargaming tasks must precede the application of AI. HCI technologies can and should be employed to gather data on discourse and decisionmaking to support AI development.

3. **AI capabilities should be employed in strategic studies to support future wargaming efforts more generally** and to shift items from the possible to the feasible. These studies include scenario generation and case identification to find challenging conditions that merit the attention of games, as well as sentiment or stance analysis in support of qualitative research on wargames.

Contents

Figures and Table

Figures

Table

Chapter 1

Introduction

Overview

The Department of the Air Force has become increasingly interested in the potential for artificial intelligence (AI) to revolutionize different aspects of warfighting. For this project, the U.S. Air Force (USAF) asked RAND Corporation researchers to consider broadly what AI can and *cannot* do in order to understand the limits of AI for warfighting applications. To address this request, we investigated the applicability of AI to four specific warfighting applications: *cybersecurity, predictive maintenance, wargames,* and *mission planning.* This report discusses the application of AI to wargames.

For the purposes of this study, we echo prior RAND reports that define *AI* broadly as "the use of computers to carry out tasks that previously required human intelligence."[1] Although the companion reports in this series emphasize the potential to use current machine-learning (ML) methods—primarily neural networks—to enable computers to perform such tasks as machine vision and natural language processing, this report embraces a broader definition that encompasses such methods as knowledge engineering. However, our analysis considers only AI technologies that exist at the time of this writing. We find that certain wargaming tasks are likely to require forms of AI that are not currently available. Although these forms of AI could become possible in the future, we can only speculate about their possible characteristics. However, other wargaming tasks are amenable to existing AI technologies.

For this analysis, we drew on three broad sources. The first was a review of historical and contemporary literature to investigate how AI has been applied for wargaming. Second, we drew on discussions with several subject-matter experts (SMEs), both inside and outside RAND, to help assess the technical feasibility and cost-effectiveness of potential AI applications. These SMEs included wargamers and AI practitioners. Finally, when necessary, we worked deductively from theoretical considerations. This was particularly important to our assessment of the prospective applicability of AI to wargaming tasks for which there have been no identifiable attempts to automate to date.

This chapter provides historical context for the interest in the use of AI in wargames. Chapter 2 presents the taxonomy of wargames that we will use to frame our analysis. Chapter 3 describes the mechanics of wargames within this taxonomy. And finally, Chapter 4 assesses the technical feasibility and cost-effectiveness of AI applications for all aspects of wargames.

---

[1] Lance Menthe, Dahlia Anne Goldfeld, Abbie Tingstad, Sherrill Lingel, Edward Geist, Donald Brunk, Amanda Wicker, Sarah Lovell, Balys Gintautas, Anne Stickells, and Amado Cordova, *Technology Innovation and the Future of Air Force Intelligence Analysis:* Vol. 2, *Technical Analysis and Supporting Material,* RAND Corporation, RR-A341-2, 2021, p. 46. This definition is updated from Marvin Minksy's original definition (Marvin Minsky, ed., *Semantic Information Processing,* MIT Press, 1968, p. v).

What Games and Artificial Intelligence Are Good For

In fiction, AI and wargames make a dangerously powerful combination. Films, television shows, and video games are populated with computers that employ AI to "game" out courses of action, allowing them to identify superior strategies. Yet, despite perennial attempts over the past few decades to make such computers a reality, they remain mostly science fiction, despite dramatic progress in ML. Why have wargaming computers failed to live up to these hopes (or to actualize the nightmares)? To understand whether and how this time might be different, it is instructive to survey the history of these attempts and how they fed into the pop culture portrayals that inform many expectations.

High Hopes and High Anxieties

The founders of AI had high expectations for game-playing algorithms and their potential applicability to military problems. In 1948, Alan Turing and his colleague David Champernowne designed the first computer chess program, Turochamp. Although Turing's attempts to implement this program on the computers of the time failed because of the limitations of those primitive vacuum-tube machines, several matches played by simulating the algorithm on paper demonstrated that the program would play valid chess. In 1950, Turing included a chess problem in his seminal paper, "Computing Machinery and Intelligence," which introduced the eponymous "Turing test" to determine whether a machine was intelligent.[2] Like many theorists of his generation, Turing intuited that playing games like chess was a quintessentially cognitive activity and that playing such games well would require a computer to have genuine intelligence, akin to that of humans.

That same year, the father of information theory, Claude Shannon, published the first paper specifically about programming a computer to play chess. He wrote that "it is hoped that a satisfactory solution of this problem will act as a wedge in attacking other problems of a similar nature and of greater significance."[3] Shannon stated eight such problems, of which one was "Machines for making strategic decisions in simplified military operations."[4] According to Shannon, "The techniques developed for modern electronic and relay type computers make [such machines] not only theoretical possibilities, but in several cases worthy of serious consideration from the economic point of view," not just as a remote eventuality but as "possible developments in the immediate future."[5]

Shannon's proposal that chess-playing machines might lead directly to computers entrusted with making battlefield decisions horrified some of his contemporaries. Norbert Wiener, the father of cybernetics, protested in his 1950 book *The Human Use of Human Beings* that, "When Mr. Shannon speaks of the development of military tactics, he is not talking moonshine, but is discussing a most

---

[2] Jeroen Fokker, "The Chess Example in Turing's Mind Paper Is Really About Ambiguity," in Andrei Voronkov, ed., *Turing-100: The Alan Turing Centenary*, EPiC Series in Computing, Vol. 10, June 2012; Andrei Voronkov, ed., *Turing-100: The Alan Turing Centenary*, EPiC Series in Computing, Vol. 10, June 2012.

[3] Claude E. Shannon, "XXII. Programming a Computer for Playing Chess," *London, Edinburgh, and Dublin Philosophical Magazine and Journal of Science*, Vol. 41, No. 314, 1950, p. 256.

[4] Shannon, 1950, p. 256.

[5] Shannon, 1950, p. 256.

imminent and dangerous contingency."[6] Wiener did not contend that such machines could not be built, but he made it clear that he believed they should not.

In the context of breathless media portrayals of early digital computers as "giant brains" solving military problems, this debate between Shannon and Wiener soon turned into inspiration for science fiction writers.[7] In November 1951, *Galaxy Science Fiction* published "Self Portrait," a tale by Bernard Wolfe in which military researchers actualize Wiener's nightmare via participants in a Manhattan Project–like effort to create a computer strategist, the Electronic Military Strategy Integrator and Calculator or "Emsiac."[8] The architect of the machine describes mechanized warfare as "the most complicated game the human race has invented so far, an elaborate form of chess which uses the population of the world for pawns and the globe for a chessboard."[9] Military officials therefore planned to turn a chess-playing computer into "a military strategy machine which can digest reports from all the units on all the fronts and from moment to moment, on the basis of that steady stream of information, grind out an elastic overall strategy and dictate concrete tactical directives to all the units."[10]

Three decades later, the film *WarGames* (1983) immortalized a similar vision of game-playing computers entrusted with military strategy for a generation of moviegoers.[11] In the film, Professor Stephen Falken is a researcher working at the intersection of AI and nuclear strategy. Falken creates a game-playing program named "Joshua" that runs on a computer called the War Operations Plan Response (WOPR) located at the North American Aerospace Defense Command. WOPR is capable not just of *reinforcement learning* (using experience to improve its performance) but also of *metalearning* (using experience to improve its ability to learn) and *transfer learning* (taking experience from one domain and applying it to a qualitatively different one). The output of WOPR's endless strategic investigations is a computer war plan that, according to one of the film's characters, "the president will probably follow" in the event of war with the Soviets. Predictably, the computer gets out of control and nearly starts World War III before it experiences a revelation at the film's climax, recognizing that nuclear war is "A STRANGE GAME" in which "THE ONLY WINNING MOVE IS NOT TO PLAY."

WarGames, therefore, emerged from a long history of popular conceptions of game-playing AI and its applicability to wargaming. But at the time *WarGames* was released in 1983, the expectations of early AI researchers about game-playing algorithms had yet to be realized.[12]

---

[6] Norbert Wiener, *The Human Use of Human Beings: Cybernetics and Society*, Houghton Mifflin, 1950, p. 206.

[7] For an early popular account of computer technology legitimizing the media conception of computers as giant brains, see Edmund C. Berkeley, *Giant Brains; or, Machines That Think*, John Wiley & Sons, 1949.

[8] Bernard Wolfe, "Self Portrait," *Galaxy Science Fiction*, Vol. 3, No. 2, November 1951. A year later, Wolfe published a novel, *Limbo*, set in 1990 after an abortive nuclear war directed by the kind of military strategy computers described in "Self Portrait" (Bernard Wolfe, *Limbo*, Hachette, 1952).

[9] Wolfe, 1951, p. 71.

[10] Wolfe, 1951, p. 71.

[11] *WarGames*, dir. John Badham, United Artists, 1983.

[12] Thomas B. Allen, *War Games: The Secret World of the Creators, Players, and Policy Makers Rehearsing World War III Today*, Berkeley Books, 1989.

Historical Experience

Playing such games as chess proved to be both easier and harder than AI pioneers in the 1950s expected it to be.[13] Simply demonstrating a working program that could play a valid game of checkers or chess made a deep impression on many contemporary observers.[14] Attempts to make computers play games drove advances in early AI research: Arthur Samuel introduced the term *machine learning* to describe his landmark checkers-playing program.[15] Moreover, the creators of such programs anticipated that they would soon improve to play these games at superhuman levels. Allen Newell made a notorious prediction that a computer would beat the world's reigning chess champion by 1967.[16] But early game-playing programs improved far more slowly than these projections. In actuality, it took until 1997, when IBM's Deep Blue bested Gary Kasparov, for a computer to best the leading chess grandmaster.[17]

Deep Blue employed alpha-beta pruning, which substituted additional search to compensate for a relatively weak ability to evaluate individual board positions.[18] Although heuristic search radically constrained the number of evaluated board positions relative to naive methods, such as the basic minimax algorithm, these programs seemed to examine moves at multiple orders of magnitude beyond that of a human player.[19] Applying this approach to tasks more complicated than chess—even to structurally similar games, such as Go—proved computationally infeasible. Because of these limitations, the algorithms used to play such games as chess could not be applied directly to strategic and military problems. And unlike the fictional portrayals of "Self Portrait" and *WarGames*, the military was not all that eager to entrust computers with real-world command authority. But some demand existed for automated wargaming, and the resultant efforts are instructive because they illuminate the contrast between chess and other strategic boardgames and wargames.

In the early 1980s, the Office of Net Assessment sponsored an effort for RAND researchers to develop automated wargaming to provide an improved methodological framework for analyzing strategic forces. The resulting RAND Strategy Assessment System (RSAS) exploited then-novel conceptual modeling techniques, supplemented with structures that drew on strategic theory and decisionmaking theory to automate political-military gaming.[20] The essence of the technique was to

---

[13] AI critic Hubert Dreyfus declared in 1965 that limited progress programming digital computers to play chess proved that "there will always be games that people can win and machines cannot" (Hubert L. Dreyfus, *Alchemy and Artificial Intelligence*, RAND Corporation, P-3244, 1965, p. 67).

[14] Daniel Crevier, *AI: The Tumultuous History of the Search for Artificial Intelligence*, Basic Books, 1993, pp. 57–58.

[15] Arthur L. Samuel, "Some Studies in Machine Learning Using the Game of Checkers," *IBM Journal of Research and Development*, Vol. 3, No. 3, 1959.

[16] Herbert A Simon and Allen Newell, "Heuristic Problem Solving: The Next Advance in Operations Research," *Operations Research*, Vol. 6, No. 1, January–February 1958.

[17] Nils J. Nilsson, *The Quest for Artificial Intelligence*, Cambridge University Press, 2009, pp. 481–484.

[18] Nilsson, 2009, pp. 91–93.

[19] Nilsson, 2009, pp. 483–484.

[20] An examination of historical RAND publications about RSAS reveals an interesting inconsistency. Sometimes the RSAS was characterized as an "automated war game," while other times it was styled as a "knowledge-based simulation." In fact, the RSAS was *both* of these things (Paul K. Davis, "Knowledge-Based Simulation for Studying Issues of Nuclear Strategy," in Allan M. Din, ed., *Arms and Artificial Intelligence: Weapon and Arms Control Applications of Advanced Computing*, Oxford University

create rule-based agents to play the roles of "red" and "blue" in the game. These agents could either be used against other computer agents or face human opponents. This approach made the behavior of the computer agents explicit: a red agent, for example, could be programmed with variants representing different postulated mindsets of Soviet leaders (or U.S. leaders).[21] Other modules of the system accounted for the interaction of military forces and other features of the simulated environment. The ultimate objective of the RSAS was to provide the best of both worlds from analytical models and traditional political-military games, offering "the richness and flexibility associated with gaming and the reproducibility and transparency obtainable with analytical models."[22] The project was bullish on man-machine interaction.

Despite some remarkable accomplishments, the RSAS left a mixed legacy. As a 2022 retrospective noted, although RAND exported the RSAS to "various government offices and war colleges," as well as the Joint Staff, "continuity proved impractical" because the "full RSAS was expensive, complicated, and demanding."[23] In particular, "the AI portions of the RSAS were seldom used outside RAND except for demonstrations."[24] At the end of the Cold War, the Department of Defense shuttered the program, and to date, there have been no other attempts to build a comparable system. The most influential component of the RSAS proved to be its global combat model, which became the Joint Integrated Combat Model that has been used for the last three decades.[25] The Joint Integrated Combat Model does not involve AI, although it includes some features that are motivated by the RSAS AI agents. Also, concepts represented in the AI agents were and are still being employed in studies, sometimes with unpretentious cognitive models expressed in tables.[26] The intent has been to "skim the cream" from what was learned from the RSAS work.

The RSAS consisted of many modular components that evolved substantially over the course of the 1980s. These included the red and blue agents, as well as the scenario and combat models. The red and blue agents had modules for strategic-level decisionmaking about escalation, among other things, and modules that represented commanders with slotted scripts adapted from AI research. The simulated commanders attempted to follow a war plan by anticipating branches or asking for new strategic guidance. If desired, humans could step in partially or wholly to make decisions in the stead of the player agents, for example, "with a Blue team playing against the Red Agent in the environment

Press, 1987; Allan M. Din, ed., *Arms and Artificial Intelligence: Weapon and Arms Control Applications of Advanced Computing*, Oxford University Press, 1987).

[21] Paul K. Davis, "Some Lessons Learned from Building Red Agents in the RAND Strategy Assessment System (RSAS)," RAND Corporation, N-3003-OSD, 1989.

[22] Morlie Hammer Graubard and Carl H. Builder, *New Methods for Strategic Analysis: Automating the Wargame*, RAND Corporation, P-6763, 1982, p. 5.

[23] Paul K. Davis and Paul Bracken, "Artificial Intelligence for Wargaming and Modeling," *Journal of Defense Modeling and Simulation*, February 2022, p. 4.

[24] Davis and Bracken, 2022, p. 4.

[25] Davis and Bracken, 2022.

[26] See Paul K. Davis, "Synthetic Cognitive Modeling of Adversaries for Effects-Based Planning," *Enabling Technologies for Simulation Science VI*, Vol. 4716, July 2002. For an even simpler manifestation, see Paul K. Davis, Angela O'Mahony, Christian Curriden, and Jonathan Lamb, *Influencing Adversary States: Quelling Perfect Storms*, RAND Corporation, RR-A161-1, 2021.

created by the Scenario Agent and the Force Agent."[27] As a 1984 RAND document noted, "If the various agents could be made intelligent enough, then the result [of pitting human players versus the simulated agents] could be a powerful mechanism for exploring concepts of strategy and the implications of different capabilities."[28] But employed with solely simulated players, the RSAS was arguably a game-structured simulation rather than a wargame in the full-fledged sense, as we will elaborate on in Chapter 2 on terminological definitions.

Conclusions

Although some early machine intelligence researchers anticipated that algorithms optimized to play such games as checkers and chess could easily be adapted for wargaming, this has not worked out in practice because those games are qualitatively different from wargaming and because war is qualitatively different from recreational gaming. Chess and checkers have simple rules and straightforward objectives, while international military confrontations do not. Attempting to account for these kinds of political-military complexities made the RSAS prohibitively expensive to maintain and operate by 1990. But the AI methods that became available in subsequent decades, such as agent-based modeling and deep reinforcement learning, are qualitatively different, so more could be possible today.

---

[27] Paul K. Davis, *RAND's Experience in Applying Artificial Intelligence Techniques to Strategic-Level Military-Political War Gaming*, RAND Corporation, P-6977, 1984, p. 8.

[28] Davis, 1984, p. 8.

Chapter 2

Definitions, Taxonomy, and Game Theory

Definitions

In analysis of national security problems, games are often defined in contrast with three other approaches: modeling and simulation, workshops and seminars, and military exercises. Confusingly, each of these alternative analytic approaches is sometimes referred to as a *game*. Some of this confusion is justified because the boundaries between the four approaches can be difficult to discern in practice.

For the purposes of this analysis, we adopt the definition used in an earlier RAND study that acknowledges that games encompass a broad family of tools used for research, education, and communication.[29] *Wargames* are a subset of these tools. Our working definition requires that a wargame incorporate four interrelated characteristics.

First, wargames are *contested*: They feature an intelligent adversary whose interests conflict with the player's interests. As Bernard Brodie noted, "What matters is the *spirit* of the gaming principle, the constant reminder that in war we shall be dealing with an opponent who will react to our own moves and to whom we must react."[30] This is a key difference between wargames and workshops and seminars, which lack this element of contest. (Some seminar games do have the character of contested activity, but they lack distinct teams.) Although some models and simulations include a contested environment, not all of them do.

Second, a wargame requires a *synthetic environment* that is modified by the players' choices.[31] This feature is different from workshops and seminars, which lack a synthetic environment, and military exercises, which employ the actual physical environment (even if in an artificially constrained way). Once again, some, but not all, models and simulations incorporate such a synthetic environment.

Third, wargames require that players *experience the projected outcomes of their decisions*. As Brodie explained, "Another benefit of war games is that they force the players to consider the problem beyond the opening moves, which it is otherwise very difficult for a war planner (or any human being) to force himself to do."[32] Once again, workshops and seminars, as well as most military exercises, lack this component, but some simulations possess it.

---

[29] Elizabeth M. Bartels, *Building Better Games for National Security Policy Analysis: Towards a Social Scientific Approach*, dissertation, Pardee RAND Graduate School, RAND Corporation, RGSD-437, 2020, Chapter 1.

[30] Bernard Brodie, *The American Scientific Strategists*, RAND Corporation, RAND P-2979, 1964, p. 30.

[31] Bartels, 2020, p. 4.

[32] Brodie, 1964, p. 31.

Finally, wargames need intelligent decisionmakers (players) who engage in strategic thinking by consciously considering the choices and actions of others.[33] Not all players have to possess this sort of intelligence, but the kind of "general" intelligence exhibited by humans is necessary for some of the applications games are used for. Without this sort of intelligence, a simulation with the other three characteristics does not rise to the level of a wargame. Eventually, machines may be able to produce the requisite kind of general intelligence that will allow fully automated wargames in the comprehensive sense to become a possibility.[34] But until this kind of artificial general intelligence becomes a practical reality, we feel it would be misleading to conflate simulations with wargames.

This definition differs from historical use of the term *wargame* that also encompassed computer simulations without human players.[35] The boundary between wargames and simulations is porous, and there are instances when it is difficult to tell whether an activity is one or the other. We elected to employ a narrower definition for the sake of consistency with the other reports in this series.[36] Computer simulation of military activity is an invaluable tool and likely also a significant practical application of AI, but such simulations are qualitatively different from the wargames we discuss in this report and would require separate analysis.

Taxonomy

For the purposes of this study, we adopt a taxonomy of wargames that was developed in an earlier RAND analysis.[37] The inspiration for this taxonomy is that different kinds of wargames are used for different purposes and, therefore, require different design trade-offs. Failure to pay adequate heed to these trade-offs can result in poorly designed wargames that waste resources, produce misleading outcomes, or both. There are certainly other ways of categorizing wargames that could be more useful or more complete for other purposes. We chose this taxonomy because it is relatively simple to describe, and the purposes for which a wargame is conducted prove to have significant bearing on the opportunities and challenges for employing AI at different stages of execution; it helps illuminate the important fault lines.

Following Bartels, we divide these wargame purposes into the four broad archetypes presented in Table 2.1.

---

[33] Bartels, 2020, pp. 4–5.

[34] For an insightful discussion of how present-day AI methods fail to rise to the needed degree of generality for these purposes, see François Chollet, "On the Measure of Intelligence," ArXiv, November 5, 2019.

[35] Garry D. Brewer and Martin Shubik, *The War Game: A Critique of Military Problem Solving*, Harvard University Press, 1979, pp. 8–10.

[36] The other reports in this series employ a narrower definition of AI that emphasizes contemporary ML and neural network techniques. Characterizing broad categories of human-assisted computer simulations as AI in this analysis would be to employ very different conceptions of "artificial intelligence" in different aspects of the project.

[37] Bartels, 2020, Chapters 3 and 4.

Table 2.1. Four Gaming Archetypes

Archetype	Definition
Systems exploration	Games that are performed to draw on human expertise and to discover new ways of thinking about defining problems.
Innovation	Games that are performed to develop hypotheses regarding potential solutions to problems.
Alternative conditions	Games that are performed to test the robustness of solutions to problems by varying the initial conditions that relate to a player's capabilities, objectives, processes, or environment.
Evaluation	Games that are performed to test whether players can perform a task according to well-defined criteria.

SOURCE: Adapted from Bartels, 2020.

The first archetype is *systems exploration* wargames. These wargames are designed to investigate the larger problem and to explore the possibilities—as it were, to better understand the game board and rule book.[38] Such wargames also reveal how experts conceptualize a situation and identify points for further investigation when they disagree.[39] Although these games may identify promising courses of action or areas of investment, they are primarily concerned with developing an understanding of how to think about a strategic problem rather than assessing specific solutions. An early example is a series of RAND and USAF games from the late 1950s that were used to explore limited war and the escalatory dynamics of conflicts in South Asia, East Asia, and the Middle East.[40] A more recent example are games that explore "gray zone" conflicts: competition between and among states and non-state actors that is somewhere in the middle of the spectrum between war and peace.[41]

The second archetype is *innovation* wargames, which seek to propose new decision options outside the status quo. The goal is to think outside the box to identify prospective solutions to strategic, operational, or tactical problems. Like systems exploration games, innovation games can be open-ended and can defy formal specification. However, innovation games differ from systems exploration games in that they are *prescriptive* (i.e., seek to find a solution to a problem) rather than *descriptive* or *diagnostic* (i.e., finding the problem). A distinctive feature of innovation games is the need for criteria to evaluate whether a proposed solution to a problem is going to be successful at meeting some operational, fiscal, organizational, or other criteria, which can make determining the success of a game difficult.[42] Examples of innovation games include the Defense Advanced Research Projects Agency's (DARPA's) games to develop the operational concepts around Mosaic Warfare and earlier efforts to identify the technical and operational foundations for defeating numerically superior Soviet military

---

[38] Bartels, 2020, p. 63.

[39] Bartels, 2020, pp. 62, 64.

[40] Milton G. Weiner, *War Gaming Methodology*, RAND Corporation, RM-2413, 1959.

[41] For example, see Becca Wasser, Jenny Oberholtzer, Stacie L. Pettyjohn, and William Mackenzie, *Gaming Gray Zone Tactics: Design Considerations for a Structured Strategic Game*, RAND Corporation, RR-2915-A, 2019.

[42] Bartels, 2020, pp. 63, 66.

forces through command and control, intelligence, and precision munitions identified through the Assault Breaker program.[43]

The remaining two wargaming archetypes are aligned with empirical research. *Alternative conditions* wargames examine how decisionmaking does or does not change in response to variations in initial conditions (i.e., it serves to check the robustness of a given strategy). It asks, How does A versus B (e.g., the availability of a certain capability) affect the decisions made by players? Obviously, this type of game must be run repeatedly with the different configurations to fulfill its purpose. Early examples are the USAF Strategy and Force Evaluation Games from the early 1960s, which sought to understand the relationship between different strategic objectives and the resulting force postures that military planners developed.[44] More recently, RAND conducted wargames for the Department of Defense that examined how different analytic products altered decisions regarding force structure designs and investments.[45] This was inspired by an earlier generation of games called Day After, which placed decisionmakers in scenarios in which a failure had occurred. The decisionmakers were then were asked to identity what they would have done to prevent the failure.[46]

Finally, *evaluation war* games aim to judge the outcomes of player decisions against some normative standard. Although every wargame involves evaluation in some form, this category of games refers to the use of wargames as part of a standardized or formalized evaluation process. An example of such a game would be a simulation that tested whether a player could succeed at some task in the face of an intelligent opponent (e.g., whether blue can prevent red from seizing an objective despite red's best effort). These games can be the simplest to run and adjudicate, but they must be well designed or they risk producing misleading results.[47] An example would be the DARPA and Center for Naval Analyses Scud Hunt wargame, which was developed to evaluate the abilities of specific force capabilities and doctrine to find and destroy mobile missiles.[48]

Importantly, the wargaming community is pragmatic, and many games could address more than one purpose. Thus, these archetypes are not mutually exclusive in practice, although they serve as a useful guide for designing and evaluating wargames by focusing attention on what sponsors, designers, and players should expect from the results. A single game can combine some or all of the features of two or more of the archetypes. However, a game that tries to do too much is likely to do none of it well because the different archetypes are in tension with each other. The comparative rigidity that

---

[43] RAND later conducted modeling and simulation work to contextualize the results of these games. See Timothy R. Gulden, Jonathan Lamb, Jeff Hagen, and Nicholas A. O'Donoughue, *Modeling Rapidly Composable, Heterogeneous, and Fractionated Forces: Findings on Mosaic Warfare from an Agent-Based Model*, RAND Corporation, RR-4396-OSD, 2021; DARPA, "Assault Breaker," webpage, undated-a.

[44] Thomas A. Brown and Edwin W. Paxson, *A Retrospective Look at Some Strategy and Force Evaluation Games*, RAND Corporation, R-1619-PR, 1975.

[45] Elizabeth M. Bartels, Igor Mikolic-Torreira, Steven W. Popper, and Joel B. Predd, *Do Differing Analyses Change the Decision? Using a Game to Assess Whether Differing Analytic Approaches Improve Decisionmaking*, RAND Corporation, RR-2735-RC, 2019.

[46] Marc Dean Millot, Roger C. Molander, and Peter A. Wilson, *"The Day After . . ." Study: Nuclear Proliferation in the Post–Cold War World—Volume II, Main Report*, RAND Corporation, MR-253-AF, 1993.

[47] Bartels, 2020, pp. 63, 66.

[48] Peter P. Perla, Michael Markowitz, Albert Nofi, Christopher Weuve, Julia Loughran, and Marcy Stahl, *Gaming and Shared Situation Awareness*, Center for Naval Analyses, November 2000.

makes alternative conditions and evaluation games useful would spoil the ability of a game to serve the purposes of innovation and possibly systems evaluation.

Game Theory and Wargaming

The attraction of applying AI to military wargames, as noted previously, is grounded in an intuition that AI technologies that can successfully play games, often with superhuman performance, will reveal insights into strategy and operations that would otherwise be inaccessible to human players. However, this requires that the games be rendered formally, as mathematical or computational structures with which players interact. Thus, the application of AI to games is implicitly related to algorithmic approaches to solving game theoretic problems.[49] Although the wargames used by military organizations differ from the formal games analyzed by game theorists, it is useful to understand their commonalities and differences to situate game-playing AI in the broader landscape of research and technology.

Wargames and game theory depart in many significant ways, and these differences will be important to consider for any application of AI to military wargaming. However, there is a common core interest that connects the games and wargames played by humans and the study of games as mathematical objects. This connection is the notion of strategic interaction, specifically the idea that when decisionmakers are interdependent, outcomes result from the combination of choices that each decisionmaker has made.[50] Thus, one actor is unable to determine their own fate; therefore, they have an interest in understanding, anticipating, and shaping the decisions of others.

Placing wargaming, gaming more broadly, and game theory on common footing with regard to the core considerations of strategic or interdependent interaction offers a clear look at how and why they diverge as research methods. Specifically, human-played and human-adjudicated games provide an opportunity to explore and evaluate strategic choices in various scenarios of interest by tapping into the expertise and experience found in individual and collective decisionmaking. In contrast, games that are primarily explored through formal means (e.g., mathematical and computational modeling and simulation) enable the identification of strategies with provably desirable properties, in some cases achieving optimality and in others, the provision of verifiable proof of an improvement over a prior standard.

As a computational technology, the potential applications of AI to contemporary and future wargames will be subject to the broader historical, institutional, and methodological challenges that military analysts and wargamers have wrestled with for decades. Although not the direct subject of this study, numerous investigations into wargaming and its applications to military and national security issues have noted that mathematical models, when applied to these issues, offer opportunities to clarify questions, isolate the importance of key variables, illuminate deep-seated structures, and

---

[49] Justin Grana, "Difficulties in Analyzing Strategic Interaction: Quantifying Complexity," in Aaron B. Frank and Elizabeth M. Bartels, eds., *Adaptive Engagement for Undergoverned Spaces: Concepts, Challenges, and Prospects for New Approaches*, RAND Corporation, RR-A1275-1, 2022; Aaron B. Frank and Elizabeth M. Bartels, eds., *Adaptive Engagement for Undergoverned Spaces: Concepts, Challenges, and Prospects for New Approaches*, RAND Corporation, RR-A1275-1, 2022.

[50] Martin J. Osborne, *An Introduction to Game Theory*, Oxford University Press, 2004.

more. However, these models rarely represent problems in such a way that provides insights into specific circumstances with specific actors at specific times.

A significant gap exists between the models and games that are intended to support concept development, acquisition, system engineering and architecture designs, and training and the analysis of specific operations of real-life friendly and adversarial forces in all their organizational and behavioral complexity.[51] Abstractions about warfare that enable analytic tractability and the employment of mathematics and computation are more compatible with peacetime military planning and innovation, from which planners and strategists can confront a broader variety of abstract pacing threats and opportunities compared with wartime when adversaries must be confronted during battle at specific times, in specific places, and with specific capabilities.[52] As Garry Brewer and Martin Shubik noted, "In applying mathematics to human affairs, including warfare, the ability to solve models must not be confused with the ability to formulate the correct or appropriate model."[53] When considering AI's applications to wargaming, an enduring challenge of how to translate mathematical or computational representations of strategic interaction into militarily relevant problems remains.

An examination of the advances in game-playing AI shows two distinguishing features that are particularly relevant to military wargaming.

First, AI research on game-playing serves a different purpose than most applications of wargaming used within military organizations and operations. Although these differences do not make the transfer of AI technologies and practices into new domains impossible, they require careful specifications of the problems and goals for which AI capabilities may be used. Specifically, games are of interest to AI researchers because they offer a basis for determining whether one algorithm can outperform another on some well-specified task.[54] Importantly, the criteria of successful research is usually narrowly defined by (1) the ability of an algorithm to outperform another algorithm or some other benchmark (e.g., a human expert) on a common task (games are useful in this regard because they produce clear winners and losers), and (2) the ability to demonstrate that an algorithm's learning processes are transferable to multiple problems and can be generalized to multiple classes of

[51] Ben Connable, Michael J. McNerney, William Marcellino, Aaron B. Frank, Henry Hargrove, Marek N. Posard, S. Rebecca Zimmerman, Natasha Lander, Jasen J. Castillo, and James Sladden, *Will to Fight: Analyzing, Modeling, and Simulating the Will to Fight of Military Units*, RAND Corporation, RR-2341-A, 2018; Elizabeth M. Bartels, Aaron B. Frank, Jasmin Léveillé, Timothy Marler, and Yuna Huh Wong, "Gaming Undergoverned Spaces: Emerging Approaches for Complex National Security Policy Problems," in Aaron B. Frank and Elizabeth M. Bartels, eds., *Adaptive Engagement for Undergoverned Spaces: Concepts, Challenges, and Prospects for New Approaches*, RAND Corporation, RR-A1275-1, 2022; Aaron B. Frank and Elizabeth M. Bartels, *Adaptive Engagement for Undergoverned Spaces: Concepts, Challenges, and Prospects for New Approaches*, RAND Corporation, RR-A1275-1, 2022.

[52] Stephen Peter Rosen, *Winning the Next War: Innovation and the Modern Military*, Cornell University Press, 1994; Barry Watts and Williamson Murray, "Military Innovation in Peacetime," in Williamson Murray and Allan R. Millett, eds., *Military Innovation in the Interwar Period*, Cambridge University Press, 1998; Williamson Murray and Allan R. Millett, eds., *Military Innovation in the Interwar Period*, Cambridge University Press, 1998; Williamson Murray and MacGregor Knox, "Conclusion: The Future Behind Us," in MacGregor Knox and Williamson Murray, eds., *The Dynamics of Military Revolution, 1300–2050*, Cambridge University Press, 2001; MacGregor Knox and Williamson Murray, eds., *The Dynamics of Military Revolution, 1300–2050*, Cambridge University Press, 2001.

[53] Brewer and Shubik, 1979, p. 78.

[54] Chollet, 2019.

problems.[55] Absent from these criteria is the development of generalizable insights into the characteristics of specific problems or games. For example, the combination of reinforcement learning (RL) and the Monte Carlo tree search (MCTS) has proven to be especially powerful for learning complex, competitive games, such as chess, Go, StarCraft II, and Defense of the Ancients, and for developing policies that defeat the world's best players.[56] However, these algorithms have not identified generalizable strategic insights into the nature of the competitive games that would allow AI capabilities to excel at a game that had never been encountered before. The advances in AI-playing games appear as generalizable approaches to learning, but not generalizable strategies from learning.

Second, advances in AI gaming have occurred in a small section of the overall game design space, most notably the region of games for which algorithmic play is possible. Many wargames overlap with that region of games and could benefit from the transfer of AI technologies into human-played games. However, a broad swath of wargames remains outside the set of state-of-the-art AI applications. For example, AI systems have excelled at games with very large state or action spaces, such as chess, Go, or StarCraft II, beating the world's best players in these contests. However, no AI system has solved these games—meaning that no algorithm has been found that can determine whether a game will end in a win, loss, or tie if both players play perfectly from the first move.[57] Indeed, the succession of improvements from one AI system to the next (e.g., AlphaGo followed by AlphaZero) demonstrates a hill-climbing process that is scaling toward an unknown fitness peak and not the achievement of a global maxima or optimum.[58] In addition to the challenges posed by state space size, other considerations include simultaneous versus sequential play, hidden or private information, asymmetric preferences, multiple players, general sum payoffs, adaptive payoffs, learning, and more. In short, as impressive as AI's achievements are, they have touched on only a small number of features that are common to military wargames and warfare.[59]

---

[55] George Skaff Elias, Richard Garfield, and K. Robert Gutschera, *Characteristics of Games*, MIT Press, 2012; Aske Plaat, Walter Kosters, and Jaap van den Herik, eds., *Computers and Games: 9th International Conference, CG 2016, Leiden, The Netherlands, June 29–July 1, 2016, Revised Selected Papers*, Vol. 10068, Springer, 2016.

[56] The term *policy* is used here deliberately and stands in stark contrast to *strategy* as it is used in the military and national security community. In AI, a policy is a mapping between a state of the world or input signal and a response to that state or signal. Learning this mapping may provide a machine with the ability to play a game at the highest competitive levels, but it does not imply that the AI has discovered a strategy that represents a deep insight into the characteristics of strategic interaction, in which inputs and outputs are joined by a set of principles that can be applied to novel situations. See Richard S. Sutton and Andrew G. Barto, *Reinforcement Learning: An Introduction*, 1st ed., MIT Press, 1998.

[57] Jonathan Schaeffer, Neil Burch, Yngvi Björnsson, Akihiro Kishimoto, Martin Müller, Robert Lake, Paul Lu, and Steve Sutphen, "Checkers Is Solved," *Science*, Vol. 317, No. 5844, September 14, 2017.

[58] David Silver, Thomas Hubert, Julian Schrittwieser, Ioannis Antonoglou, Matthew Lai, Arthur Guez, Marc Lanctot, Laurent Sifre, Dharshan Kumaran, Thore Graepel, Timothy Lillicrap, Karen Simonyan, and Demis Hassabis, "A General Reinforcement Learning Algorithm That Masters Chess, Shogi, and Go Through Self-Play," *Science*, Vol. 362, No. 6419, December 7, 2018.

[59] Examples of games that display these features, such as Dungeons and Dragons, are discussed later in the report.

Chapter 3

The Mechanics of Wargames

Many military wargames follow a typical progression. First, the participants are divided into opposing teams and are provided with a description of the game scenario, including specific information on their team's objectives and capabilities. Second, the participants plan their actions and submit their decisions or moves to the wargame's control cell. Third, the control cell evaluates the combination of moves, determines the outcomes, and briefs the results back to the participants; this process may be repeated several times over the course of wargame. Fourth, observations of the participants' decisionmaking, decisions, and outcomes are aggregated for analysis and assessment, and there is often a discussion about the wargame held with all the participants, control, and observers.[60]

This can be a useful way to think of wargaming activity. For our purposes, however, a more canonical description of wargame phases is more useful to consider how AI technologies could be inserted into the conduct of wargames. The four broad areas or phases of wargaming activity are: preparation, play, adjudication, and interpretation.[61] Although at a high level, many of these activities occur in multiple phases or as subcycles within cycles of gaming. For example, preparing a wargame will involve not only the development of the game's scenario and playing mechanics but also play-testing, which includes an assessment of adjudication capabilities and strategies, data collection, and an interpretation of frameworks from which analytic findings and broader assessments may be based. In the following sections, we discuss each phase and how it may differ for games conducted for different purposes, as described in Chapter 2.

Preparation

AI technologies may be successfully employed to prepare games when the specific scenarios, force structures, and mechanics are well specified. In most cases, game preparation is a costly, but largely manual, effort with limited opportunities for AI applications. However, there are certain bespoke applications for AI that may be both possible and useful. Those examples include

- automatically updating the order of battle or scenario information from external databases and interpolating missing data required for gameplay

---

[60] This list is derived from Sherrill Lingel, Jeff Hagen, Eric Hastings, Mary Lee, Matthew Sargent, Matthew Walsh, Li Ang Zhang, and David Blancett, *Joint All-Domain Command and Control for Modern Warfare: An Analytic Framework for Identifying and Developing Artificial Intelligence Applications*, RAND Corporation, RR-4408/1-AF, 2020.

[61] Bartels, 2020.

- procedurally generating gaming environments, such as geographic terrain, urban environments, maps, and interstate systems[62]
- creating simple game-playing agents that could be used to provide players with interactive tutorials regarding game mechanics and objectives
- creating or completing scenario details and background materials that provide players with narrative detail regarding the game environment and player motives.[63]

Although all game types have the potential to use AI in these capacities during preparation, it should be noted that innovation and systems exploration applications are less likely to pose questions that are as well specified, making their use less likely.

More-complex and more-ambitious applications on the frontier of AI would include the design of game scenarios and the automated discovery of interesting or challenging scenarios for which human attention is best used by playing manually. Applications of this kind would require situating wargames within a broader campaign of learning that would combine games, modeling and simulation, exercises, and historical case studies into a unified analytic effort in which the outputs of one approach inform the focus and design of the next.[64] The feasibility of AI applications in such a capacity is necessarily contingent on the presence of modeling and simulation capabilities that can identify interesting or challenging strategic, operational, or tactical scenarios that would merit human attention.

Play

Advancements in AI gameplay are among the most-exciting technological developments of the ongoing second wave of AI systems.[65] Each of the four game purposes could have different applications and different levels of difficulty in the employment of AI to the role of players in wargames.

Systems Exploration Games

AI gameplay may be possible, although limited, in systems exploration games. Gameplay is possible in this case largely because of the relatively relaxed and open definitions of meaningful results. Specifically, a machine that is capable of generating novel and unexpected behavior can produce outcomes that may be stimulating and aid in the generation and exploration of potential futures. In

---

[62] For examples, see Jakub Wabiński and Albina Mościcka, "Automatic (Tactile) Map Generation—A Systematic Literature Review," *ISPRS International Journal of Geo-Information*, Vol. 8, No. 7, 2019; George Kelly and Hugh McCabe, "A Survey of Procedural Techniques for City Generation," *ITB Journal*, Vol. 7, No. 2, 2006; and Mingyun Wen, Jisun Park, and Kyungeun Cho, "A Scenario Generation Pipeline for Autonomous Vehicle Simulators," *Human-centric Computing and Information Sciences*, Vol. 10, 2020, p. 24.

[63] Debjit Paul and Anette Frank, "COINS: Dynamically Generating COntextualized Inference Rules for Narrative Story Completion," arXiv, June 4, 2021.

[64] Paul K. Davis, "Illustrating a Model-Game-Model Paradigm for Using Human Wargames in Analysis," RAND Corporation, WR-1179, 2017.

[65] John Launchbury, "A DARPA Perspective on Artificial Intelligence," briefing slides, Defense Advanced Research Projects Agency, undated.

such circumstances, AI-based gameplay is possible because of the affordances of the research purpose to be exploratory and to search for interesting outcomes rather than because of the sophistication of the technology.

The uses of AI for gameplayers in this regard shifts the purpose of humans to engaging with one another to creatively explore how a complex system of strategically interacting competitors may work or to allowing machines to perform this generative task, with the results evaluated by humans.[66] Intuitively, this search for "interestingness" mimics how generative AI techniques (e.g., genetic algorithms) have been employed in industry to develop novel options for decisionmakers who are unsure of what they are looking for and to use the exploratory process to discover their preferences.[67] Importantly, moving human experts from a game-playing role to a game-evaluation role would challenge the strict definitions of whether the analytic approach would then qualify as a wargame.

Innovation Games

Applying AI gameplay to innovation games is the most challenging application, but it is the most attractive as a means for advancing military and national security strategy. In these applications, the desired outcomes are not only exploratory but are also intended to signal the existence of strategies, operational concepts, tactics, organizational designs, and technologies that would be worthy of further investment. Speculations regarding the use of computers for such a role to discover hidden signals about the operational characteristics of the future of warfare have simultaneously served as both a desideratum for net assessment and as a justification to keep human-played wargaming as a part of the overall defense establishment in order to maintain lines of research that traditional modeling and simulation cannot credibly perform.[68]

Thus, to successfully play innovation games, the AI must not only be independently creative but its proposed solutions to problems in the game must be located within a plausible region of the design space and not violate hard constraints.

Importantly, many AI systems have demonstrated an ability to play games at superhuman levels and defeat the best human players. Moreover, these gameplaying AI systems have identified novel and innovative strategies that have surprised human players. What distinguishes games for innovation from those that AI has demonstrated the ability to play at a master level is the level of complexity regarding the ambiguous or asymmetric payoffs or the win conditions and often non-zero sum character of the player's goals, as well as the roles of uncertainty, communication, and deception. Indeed, the largest advances in gameplay have stemmed from the ability of computers to handle increasingly large action spaces, while other sources of complexity remain problematic.[69]

---

[66] On the role of interactions as a generator of creative and unpredictable outcomes, see R. Keith Sawyer, *Social Emergence: Societies as Complex Systems*, Cambridge University Press, 2005.

[67] Icosystem, "Hunch Engine," webpage, undated.

[68] Stephen Peter Rosen, "Net Assessment as an Analytical Concept," in Andrew W. Marshall, J. J. Martin, and Henry S. Rowen, eds., *On Not Confusing Ourselves: Essays on National Security Strategy in Honor of Albert and Roberta Wohlstetter*, Westview Press, 1991; Andrew W. Marshall, J. J. Martin, and Henry S. Rowen, eds., *On Not Confusing Ourselves: Essays on National Security Strategy in Honor of Albert and Roberta Wohlstetter*, Westview Press, 1991.

[69] James Goodman, Sebastian Risi, and Simon Lucas, "AI and Wargaming," arXiv, September 25, 2020; Grana, 2022; Frank and Bartels, 2022.

Alternative Conditions Games

AI gameplay in games that are intended to explore the performance of a given strategy, operational concept, tactic, or other feature under alternative conditions may be feasible within a narrow range of search. More specifically, an AI player may be able to reveal sensitivities and insensitivities to changes in initial conditions if the underlying game is effectively playable. However, the successful application of AI technology for these purposes is contingent on the existence of formal models that can be used to develop and maintain AI players.

To illustrate these differences, consider two alternative AI systems for playing chess. One system employs expert heuristics to consider the board state and to determine what moves to make. The second system is a trained statistical system that employs contemporary data-driven ML methods, such as the combination of RL and MCTS. If an alternative version of chess were introduced—a version in which a single pawn was replaced with a third knight—the first AI system would be able to adapt and continue to play the game variant (although perhaps with a loss of skill). The second AI system, however, would require retraining, potentially at great expense in terms of time and cost, before it could adapt to play a relatively nearby game in the larger space of possible game designs. Therefore, the potential for AI to successfully play strategic games under alternative conditions depends largely on the underlying robustness of the AI systems and the presence of models that are isomorphic with the wargame's environment and action space.

Evaluation Games

Among the most realistic applications of AI for the purposes of gameplay is the use of AI players in evaluation games that are used to train and test the development of human judgment and expertise. In these cases, games may be used in a predictable fashion, situating human players in circumstances with the goal of teaching or evaluating specific skills. Because of the presence of bounded conditions, specific learning objectives, and a large number of plays of the game compared with games used for research, the possibility of training an AI for human players to interact with could be a realistic possibility. Once generated, the AI could play an important role in teaching and evaluation (e.g., by automatically adjusting its level of difficulty on the basis of how a human player performs).[70] As training becomes increasingly digitized, the opportunities to combine multiple technologies into an integrated platform that combines real and artificial assets and actors increases. Contemporary and near-term applications of AI in training, based on the use of three-dimensional modeling and simulation and extended-reality human-computer interaction (HCI), will allow for the use of AI in live, virtual, and constructed environments.[71]

---

[70] Dan Ayoub, "Unleashing the Power of AI for Education," *MIT Technology Review*, March 4, 2020.

[71] Pranshu Verma, "Fighter Pilots Will Don Augmented Reality Helmets for Training," *Washington Post*, August 4, 2022; Joaquin Victor Tacla, "Air Force Pilots Will Use AR Headsets to Fight AI-Powered Enemies!" Tech Times, August 5, 2022.

Adjudication

Systems Exploration Games

The employment of AI to adjudicate systems exploration games is a potential application of AI in a limited sense. Because systems exploration games are open-ended, a full representation of the system, all possible manipulations, and the potential to find novel configurations remains a conceptual and technical challenge.[72] Instead, AI could play a role in adjudicating player interactions in partial, limited ways. For example, a speech-to-text system combined with a large language model could be used to determine the outcome of negotiations between human players.[73] This example suggests that AI systems may not be able to completely adjudicate all aspects of human-played games but could productively assist in resolving certain classes of interaction and perform partial computations to assist human adjudicators. Notably, research into the uses of AI in open-ended discourse- and narrative-driven games, such as Dungeons and Dragons, is on the frontier of research.[74]

Innovation Games

Because the required performance of AI systems is higher for innovation games, the needed level of performance from an adjudication system is more difficult to attain than for other gaming types. The differences may be thought about in terms of scientific processes. Systems exploration games seek to generate potential futures of interest and are closely related to processes of hypothesis generation. Philosophers of science often argue that the creative process of hypothesis generation sits outside the scientific method; similarly, there are many different kinds of machinery for generating hypotheses and a high tolerance for inefficiency could allow AI systems to be productively employed in that role.[75] However, innovation games require not only that the games generate novel solutions to problems but also demand that a preliminary analysis of the solution appears plausible. In this regard, the innovation game allows for both the generation of a hypothetical solution to a problem and a weak or soft test of the solution's feasibility based on its adjudication procedures and results.[76] Although the notion that wargames provide a basis for rigorous hypothesis testing is controversial, innovation games offer a weak form of testing based on the minimum criteria of internal validity within the limits of a world constructed by the game designers, players, and control.

---

[72] Kenneth O. Stanley, Joel Lehman, and Lisa Soros, "Open-Endedness: The Last Grand Challenge You've Never Heard of," O'Reilly Media, December 19, 2017; Arend Hintze, "Open-Endedness for the Sake of Open-Endedness," *Artificial Life*, Vol. 25, No. 2, Spring 2019.

[73] Kirk Ouimet, "Conversations with GPT-3," Medium, July 16, 2020.

[74] Beth Singler, "Dungeons and Dragons, Not Chess and Go: Why AI Needs Roleplay," Aeon, April 3, 2018; David Cassel, "Can We Teach an AI to Play Dungeons and Dragons?" The New Stack, March 28, 2021.

[75] Patrick W. Langley, Herbert A. Simon, Gary Bradshaw, and Jan M. Zytkow, *Scientific Discovery: Computational Explorations of the Creative Process*, MIT Press, 1987, pp. 3–36.

[76] The notion of a weak test is one that possesses *face validity*, meaning that the results are intuitively plausible to experts, as opposed to a hard test that would require independent, rigorous, and specialized methods to administer. Although games can often provide the former, the latter is a task that games cannot perform because of the artificial nature of their design, requiring the use of other methods to complete. See Bartels, 2020.

Given these criteria, the technical feasibility of employing AI in the adjudication of innovation games is among the most difficult of use cases for the game types that exist.

Alternative Conditions Games

The employment of AI in the adjudication of games that explore alternative conditions is largely feasible for classes of games in which formal models play an important role in adjudication processes. In wargames that are designed to explore alternative conditions of force structures, capabilities, and behaviors, the major source of variation exists within the game's initial conditions but not at the level of altering player goals or the causal model of the world. Therefore, the game's internal model of the world is largely held constant across different input conditions.

In such a framing, the possibility of using AI in adjudication processes is possible, although there are broadly two alternative implementation strategies to consider in game development and performance.

First, adjudication models can be developed and trained prior to gameplay. These may be partial models that create "lookup tables" of outcomes for specific interactions that can be used by the control team to resolve specialized parts of the overall game. This is possible when the action space of the game is well-defined and constrained and when resolution via simulation is possible, which allows for RL systems to compute likely or best-response outcomes to player interactions. For most games that seek to employ contemporary and emerging AI capabilities, a computational model of the complete game or a component of the game is necessary to allow AI systems to be trained in advance. Given the time and expense required to train these systems, this would need to be performed in advance of the game itself.

Second, more-classical versions of AI that employ rules-based behaviors (e.g., expert systems or agent-based models) may be employed in real-time gameplay. These systems may respond to player's choices in real time and aid in determining whether the player's actions succeed or fail in meeting their major or minor objectives. For example, psychological models may be used to determine whether military forces will obey orders and knowingly move into harm's way if ordered to by the players commanding them.[77] Applications of this type effectively turn computational models into the gameboard or rulebook, blurring the line between AI agents that serve as additional players and those that resolve the actions of human gameplayers.[78]

Evaluation Games

The use of AI in the adjudication of evaluation games has already become commonplace. For example, human players have interacted with computational agents in training exercises for years, in which the agents, essentially models, automatically evaluate and respond to the actions of the players. At higher levels, models of complex social systems have also been used in games that train and evaluate gameplayers. A notable example was the use of economic simulations to adjudicate the

---

[77] Connable et al., 2018.

[78] Bartels et al., 2022.

macroeconomic policy decisions of gameplayers from postcommunist states transitioning to free market economies.[79] Although these kinds of models would not qualify as contemporary AI systems, advances in computational social sciences that construct dynamic models from empirical data provide opportunities to create new generations of interactive models fulfilling the same purpose.[80] Indeed, contemporary AI systems and future developments, particularly those that expand the realism of computational models of military and strategic systems, the speed of their performance, and the ways in which human players can interact with them, are likely to increase in availability, expanding the frontier of AI technology applications to wargames in these purposes.

Interpretation

Systems Exploration Games

The purpose of systems exploration games is to discover potential futures of interest based on the interactions between players (and control). AI provides many opportunities to find emergent patterns and structures that may be of interest, mostly by drawing on a body of data analytic techniques that are referred to as *unsupervised learning methods*. These techniques include principal components analysis, K-means clustering, and many other exploratory data analysis methods.[81]

Although these techniques can be used to analyze complex data that span different types (e.g., numerical, text, audio, video, geolocation), the identification of structure and patterns does not automatically produce any meaning that analysts and decisionmakers can understand. Explainable AI is a research frontier that is designed to make the processes and outputs of AI systems interpretable by the users of their systems.[82]

The challenges posed by explanation are further exacerbated by the need for systems to understand what is discovered within the data. Although AI systems may have produced truly remarkable results performing specific tasks, they have not yet demonstrated the ability to understand the meaning of those tasks, much less the appropriateness of what tasks to perform in complex and changing environments.[83] For example, the current generation of large language models have generated truly impressive text, mirroring the prose of very sophisticated authors.[84] However, these systems do

---

[79] Richard H. White, Edward Smith, An-Jen Tai, William E. Cralley, Joel Christenson, David Davis, William Fedorochko, Jr., Jack LeCuyer, Dayton Maxwell, Klaus Niemeyer, Edwin Pechous, Danielle Philips, Ian Rehmert, Marc Samuels, and Katherine Ward, *An Introduction to IDA's S.E.N.S.E.—R.S.A. Project*, Institute for Defense Analyses, September 1999.

[80] Robert L. Axtell, "Short-Term Opportunities, Medium-Run Bottlenecks, and Long-Time Barriers to Progress in the Evolution of an Agent-Based Social Science," in Aaron B. Frank and Elizabeth M. Bartels, eds., *Adaptive Engagement for Undergoverned Spaces: Concepts, Challenges, and Prospects for New Approaches*, RAND Corporation, RR-A1275-1, 2022; Frank and Bartels, 2022.

[81] Nicholas J. Higham, ed., *The Princeton Companion to Applied Mathematics*, Princeton University Press, 2016.

[82] For example, see Google Cloud, "Explainable AI," website, undated; Jesus Rodriguez, "Introducing AI Explainability 360: A New Toolkit to Help You Understand What Machine Learning Models Are Doing," KDnuggets, August 27, 2019; DARPA, "Explainable Artificial Intelligence (XAI) (Archived)," webpage, undated-b.

[83] Chollet, 2019.

[84] GPT-3, "A Robot Wrote This Entire Article. Are You Scared Yet, Human?" *The Guardian*, September 8, 2020; Jacob Bergdahl, "An AI Wrote This Story," Medium, July 16, 2021.

not understand the meaning of the text they generate, effectively mimicking the production of narrative prose without comprehending its meaning.

These limitations constrain the application of AI to the interpretation of open-ended games, such as those used for exploring systems. AI systems could play important roles in extracting information about hypothetical systems that emerge from gameplay. However, these contributions will likely remain bounded and insufficient to replace or minimize the requirements for human experts to assess the results of games.

Innovation Games

Innovation games will face similar challenges to those of systems exploration games.

Alternative Conditions Games

The interpretation of the results of alternative condition games could be the largest beneficiaries of AI technologies. Because the analysis objective of alternative condition games is to detect similarities and differences between different starting conditions (i.e., treatments to players and their capabilities in experimental terms), AI methods that can identify and measure qualitative and quantitative differences between items may be usefully applied to comparing games. In these cases, AI systems may not need to be fully aware of the meaning of differences between games, provided it can characterize what is different and by how much.

Evaluation Games

The feasibility of using AI to adjudicate evaluation games is based on the characteristics of the solution. Because evaluation games are largely based on the teaching of or testing for specific competencies, the indicators of proficiency and the extent to which AI can identify and measure those games is deterministic. For example, games that are intended to examine how soldiers maneuver through urban terrain or coordinate air and ground forces against an entrenched adversary could employ a variety of sensors and algorithms to determine employment of the desired tactics, techniques, and procedures; concept of operations; or doctrine. AI applications could include machine vision, radar processing, and other sensing technologies, for which data are processed by algorithms to find patterns and structure in time and space to create quantitative profiles of maneuvers. There has also been significant success with AI generating text.[85] By comparison, AI systems for *evaluating* the writing of reports remain difficult for current natural language understanding systems, and higher-

---

[85] The results appear to be reasonably well written but can lack meaningful content: They have been described as "fluent but not factual." See Chris Stokel-Walker and Richard Van Noorden, "What ChatGPT and Generative AI Mean for Science," *Nature*, February 6, 2023.

order evaluation tasks, such as determining the replicability of scientific research or the quality and strength of analytic reasoning, remain on the frontiers of advanced research.[86]

---

[86] Elaine Lin Wang, Lindsay Clare Matsumura, Diane Litman, Richard Correnti, Haoran Zhang, Zahra Rahimi, Zahid Kisa, Ahmed Magooda, Emily Howe, and Rafael Quintana, *Contributions to Research on Automated Writing Scoring and Feedback Systems*, RAND Corporation, RB-A1062-1, 2022; Nazanin Alipourfard, Beatrix Arendt, Daniel Benjamin, Noam Benkler, Mark Burstein, Martin Bush, James Caverlee, Yiling Chen, Chae Clark, Anna Dreber, Timothy M. Errington, Fiona Fidler, Nicholas Fox, Aaron Frank, Hannah Fraser, Scott Friedman, Ben Gelman, James Gentile, C. Lee Giles, Michael Gordon, Reed Gordon-Sarney, Christopher Griffin, Timothy Gulden, Krystal Hahn, Robert Hartman, Felix Holzmeister, Xia Hu, Magnus Johannesson, Lee Kezar, Melissa Kline Struhl, Ugur Kuter, Anthony Kwasnica, Dong-Ho Lee, Kristina Lerman, Yang Liu, Zach Loomas, Bri Luis, Ian Magnusson, Michael Bishop, Olivia Miske, Fallon Mody, Fred Morstatter, Brian A. Nosek, E. Simon Parsons, David Pennock, Thomas Pfeiffer, Haochen Pi, Jay Pujara, Sarah Rajtmajer, Xiang Ren, Abel Salinas, Ravi Selvam, Frank Shipman, Priya Silverstein, Amber Sprenger, Anna Squicciarini, Stephen Stratman, Kexuan Sun, Saatvik Tikoo, Charles R. Twardy, Andrew Tyner, Domenico Viganola, Juntao Wang, David Wilkinson, Bonnie Wintle, and Jian Wu, "Systematizing Confidence in Open Research and Evidence (SCORE)," Center for Open Science, May 2021; Steven Rieber, "Methods for Evaluating Analytic Quality of Reasoning RFI Number IARPA-RFI-22-02," Intelligence Advanced Research Projects Activity, November 30, 2021.

Chapter 4

Cost-Benefit Analysis of Artificial Intelligence Wargaming Applications

Assessment Measures

For the employment of AI to be of practical benefit for a wargaming task, it has to be both *technically feasible* and *cost-effective*. For the purposes of this analysis, we define *technically feasible* as possible using available technology or technology that appears likely to exist in the foreseeable future (within the next decade). According to this definition, an AI application that requires some kind of technical breakthrough, even a relatively minor one, is classed as infeasible. Although in some cases this could result in overly conservative conclusions, we believe that it is preferable to err on the side of caution rather than unjustified optimism where unpredictable technological progress is concerned.

Furthermore, some AI applications that are theoretically possible using present-day methods are still infeasible in practice because of nontechnical constraints. For instance, given sufficient training data, state-of-the-art ML methods can address a wide array of daunting and ill-specified problems.[87] In principle, these methods could be applied to difficult wargaming tasks just as they have been to such games as StarCraft and such tasks as text and image synthesis.[88] But to attain useful performance, these methods generally require quantities of training data that we cannot acquire for wargaming tasks at any cost; the digitization of games is required first. An example would be a wargame about unfamiliar weapon systems interacting in an unfamiliar environment. The problem is that the desired knowledge simply does not exist to either be translated into training data or used to construct a simulator that could then serve as a training environment for a reinforcement learning agent. In the future, ML methods with higher sample efficiency could be developed, which would make it feasible to use AI for a broader array of wargaming tasks; however, the emergence of such a method would constitute a major breakthrough that the USAF should not count on for planning purposes.

---

[87] Scott Reed, Konrad Żolna, Emilio Parisotto, Sergio Gómez Colmenarejo, Alexander Novikov, Gabriel Barth-Maron, Mai Giménez, Yury Sulsky, Jackie Kay, Jost Tobias Springenberg, Tom Eccles, Jake Bruce, Ali Razavi, Ashley Edwards, Nicolas Heess, Yutian Chen, Raia Hadsell, Oriol Vinyals, Mahyar Bordbar, and Nando de Freitas, "A Generalist Agent," *Transactions on Machine Learning Research*, November 2022.

[88] Oriol Vinyals, Igor Babuschkin, Wojciech M. Czarnecki, Michaël Mathieu, Andrew Dudzik, Junyoung Chung, David H. Choi, Richard Powell, Timo Ewalds, Petko Georgiev, Junhyuk Oh, Dan Horgan, Manuel Kroiss, Ivo Danihelka, Aja Huang, Laurent Sifre, Trevor Cai, John P. Agapiou, Max Jaderberg, Alexander S. Vezhnevets, Rémi Leblond, Tobias Pohlen, Valentin Dalibard, David Budden, Yury Sulsky, James Molloy, Tom L. Paine, Caglar Gulcehre, Ziyu Wang, Tobias Pfaff, Yuhuai Wu, Roman Ring, Dani Yogatama, Dario Wünsch, Katrina McKinney, Oliver Smith, Tom Schaul, Timothy Lillicrap, Koray Kavukcuoglu, Demis Hassabis, Chris Apps, and David Silver, "Grandmaster Level in StarCraft II Using Multi-Agent Reinforcement Learning," *Nature*, Vol. 575, November 14, 2019; Aditya Ramesh, Mikhail Pavlov, Gabriel Goh, Scott Gray, Chelsea Voss, Alec Radford, Mark Chen, and Ilya Sutskever, "Zero-Shot Text-to-Image Generation," *Proceedings of the 38th International Conference on Machine Learning*, Proceedings of Machine Learning Research, Vol. 139, 2021.

To be of practical benefit for the USAF, an AI application for wargaming needs to be not just feasible but also cost-effective. This could impose a much higher barrier to AI adoption for these purposes, subject to budgetary limitations. To continue the previous example, perhaps the needed data to build the system are available but only at an extremely high cost, such as hiring large numbers of human players to play a game many times to build up the needed training set.[89] In this instance, the AI agent only becomes cost-effective when the resulting model is used an immense number of times because getting the training data demands playing the game more times than might be needed to play a conventional manual game to answer a question of interest.

Additionally, data costs are far from the sole costs of applying AI to wargaming tasks. Skilled talent to build and maintain AI models is a notorious bottleneck in the field, with qualified practitioners sometimes commanding seven-figure salaries.[90] Moreover, many AI tools require skilled users and builders—just interpreting the model output may demand considerable and scarce human talent. Last, but by no means least, are the hardware and other capital costs that are required to build and run a model. Large state-of-the-art ML models, such as those developed by research leaders OpenAI and DeepMind, often require millions of dollars' worth of computer time to train the models, independent of data and labor costs.[91] Until these costs can be reduced (most likely from the introduction of more sample-efficient ML methods), many applications of AI to wargaming will not be justifiable on the basis of cost, even though they are within the realm of present-day technological possibility.

Finally, to be cost-effective, an AI application has to both be affordable and offer some meaningful benefit over available alternatives that justifies the cost. In some cases, AI will enable new possibilities that were simply unavailable with traditional methods. But an AI application that simply replicates the output of older approaches can be hard to justify on a cost-benefit basis: Even if the cost is relatively modest, the marginal benefit is too slim. Because the purpose of this report is to assess the limits of AI in warfighting, we have set a high bar for AI and marked technical feasibility accordingly. However, in cases when human-machine teaming might succeed when AI alone cannot, we have added caveats to indicate when certain cases could be handled in that manner.

Analysis

We drew on three broad sources for our analysis. The first was a literature review of both historical and contemporary publications to investigate how AI has been applied to wargaming. The second was a series of discussions with SMEs, including wargamers and AI practitioners. Finally, when necessary, we worked deductively from theoretical considerations when there were no identifiable attempts to date to automate the particular wargaming tasks in question. Figure 4.1

---

[89] This would be the gaming equivalent of the approach used by Project Maven, which employed humans to identify objects of interest in images to build up a training set for image recognition applications.

[90] Diana Gehlhaus, Luke Koslosky, Kayla Goode, and Claire Perkins, *U.S. AI Workforce: Policy Recommendations*, Center for Security and Emerging Technology, October 2021.

[91] Andrew J. Lohn and Micah Musser, *AI and Compute: How Much Longer Can Computing Power Drive Artificial Intelligence Progress?* Center for Security and Emerging Technology, January 2022.

summarizes our estimates of the *technical feasibility* of employing AI for wargaming by task and purpose.

Figure 4.1. Technical Feasibility of Artificial Intelligence for Wargaming

Type \ Task	Preparing	Playing	Adjudicating	Interpreting
Systems Exploration	Feasible for well-specified use cases	In poorly-explored domains AI agents may be useful	Feasible for well-specified use cases	Requires "human-level" AI
Innovation		Requires "human-level" AI	Possibility of unforeseen player actions	Requires "human-level" AI
Alternative Conditions		In poorly-explored domains AI agents may be useful	Feasible where player action space is known and bounded	Differences can be measured even when specification is weak
Evaluation		AI player can be adequate to show (un)reachability of normative goal	Feasible where player action space is known and bounded	May be difficult for AI to judge if goal is met in some cases

Prohibitive	Possible	Feasible

Each cell in this figure presents an assessment of the technical feasibility of employing AI methods for the game's purpose and performance. As is evident, the technical feasibility of wargaming tasks is largely determined by the purposes for which wargames are intended to serve.

When costs are taken into account, they modify Figure 4.1 largely by column or wargaming task. Consideration of the costs associated with applying AI to wargaming is largely based on three dimensions: (1) the number of times a game will be played, (2) the extent to which the game has been digitized, and (3) the openness of the game regarding player options, interactions, and adjudication. Together, these three dimensions characterize the extent to which military organizations may fruitfully invest in the development and use of technically feasible (but resource-intensive) design, development, testing, and deployment of AI to support wargaming activities. In the following sections, we describe how these three factors affect the cost-benefit analysis. We conclude with a diagram that combines technical feasibility with cost considerations.

Number of Plays

The number of times a game will be played has significant influence over two major considerations of AI development and one minor consideration. First, the number of times a game is played represents the extent to which there are observational opportunities to gather information on the decisionmaking of the players and control. This effectively sets the floor for considering the quantity of data that are available to support the development of AI systems to support the wargame. It is important to note, however, that it is not necessary to collect enough data to formally train the AI algorithms, which would be very challenging: There is a lesser range in which enough observations may inform the design of expert systems or other computational agents from observing human players.[92]

Second, the number of times a game is played helps define the market or demand for which analysis of the game will be performed. In this case, the labor savings and benefits of consistency across play (e.g., increasing the consistency of game adjudication) would be most affected by the employment of AI. Alternatively, when a game is played infrequently or even as a one-off event, the benefits of AI are limited to its contributions for a single use. In most instances, few problems would warrant the costs of developing sophisticated AI applications to support a game that will be played only a few times.

Finally, a lesser consideration regarding the number of times a game is played is related to its design and adjudication processes: The more times a game is played, the more likely its internal mechanics, such as player actions and adjudication processes, are to be codified and rule-bound. The implications are that games that are played fewer times are more likely to be open and adaptive, raising the technical level and, therefore, the expense of employing AI in their conduct.

The Extent of Digitization

The extent to which a wargame is digitized constitutes another driver of costs for developing and employing AI within a wargame.

One of the most challenging and costly aspects of wargaming is the investment that is needed to capture the players' deliberations, actions, and accumulated small choices to get a full accounting of the strategic interaction. Likewise, in games in which models are used to adjudicate the actions of players, a large expense in terms of time and labor is attributed to translating players' chosen moves into model inputs that can then be simulated by the control cell.

Alternatively, many games, particularly training games, are already played in completely digital environments. These may include computer-based games played through traditional mouse, keyboard, and joystick input devices, to sophisticated simulators that allow for interactions with real or artificial game assets in virtual, augmented, or mixed reality.

The costs of bringing AI into games will largely be contingent on the extent to which the players and control are already interacting on a digital infrastructure. In this regard, investments in HCI, a set of technologies that are laden with AI, might be a necessary precursor to create the conditions for more-advanced AI to be brought to bear. Thus, in addition to thinking about gameplaying AI agents and models for adjudication, an infrastructural foundation could exist in less obvious but necessary investments that include:

- speech-to-text technologies to capture verbal interactions between players[93]

---

[92] It is an open question whether sufficient data can be gathered to train an AI system to actually play the game, even with a high sample efficiency. However, that is only one of the many tasks an AI might perform in a wargame.

[93] Amazon Web Services, "Amazon Transcribe," website, undated.

- virtual presence technologies to allow sustained face-to-face, naturalistic participation in meetings[94]
- mixed reality to facilitate simultaneous interaction between real and artificial gaming assets (e.g., collectively marking maps, moving game pieces, or engaging in simulated combat and other missions).[95]

Importantly, in the context of gaming, each of these technologies is an investment area and must be carefully deployed and tested to ensure that (1) player behavior is not an unanticipated artifact of the technology and (2) that behavioral artifacts do not become the basis for AI systems that mine game data for lessons learned or for creating virtual agents to play or adjudicate future games.

The Openness of Games

A central theme in the analysis of gaming, its history, and uses is the tension between two broad approaches to wargames. In one version, games are bound by tightly constructed and specified rules that dictate what players can and cannot do and how their actions and interactions will be adjudicated. In the other version, games are cocreated by participants, a process in which designers, players, and adjudicators collectively and creatively develop and evaluate the actions and interactions of the players, which allows games to elicit and reveal knowledge and insights that might not have been present or explicit beforehand.

The more rigid a game's design and constrained its conduct, the more opportunities exist to bring computation to bear on the game's conduct.[96] This primarily includes the use of computational models that capture players' behaviors and adjudicates their interactions. Alternatively, the moves and the resolution of open-ended games rely more on discourse and argumentation (e.g., matrix games) that are both difficult to capture digitally and to resolve without a priori knowledge of what actions and interactions are possible.[97]

When games are bounded, as is the case for those used in training and testing (i.e., for the purposes of evaluation), the costs of developing AI to be employed in their use can be lowered because of the previously discussed two factors—it is more likely that there will be more opportunities to collect data and to employ formal models within a digital game-playing environment. Alternatively,

---

[94] Microsoft, "Holoportation," website, undated; Martijn J. Schuemie, Peter van der Straaten, Merel Krijn, and Charles A. P. G. van der Mast, "Research on Presence in Virtual Reality: A Survey," *CyberPsychology & Behavior*, Vol. 4, No. 2, July 2004; Anabela Marto and Alexandrino Gonçalves, "Augmented Reality Games and Presence: A Systematic Review," *Journal of Imaging*, Vol. 8, No. 4, April 2022, p. 91.

[95] For example, see Simtable, "About Simtable," webpage, undated; Laurence Russell, "Mirrorscape Wants to Conjure Your Favorite Tabletop Game in AR," *Wired*, January 18, 2022.

[96] The success of AI in such games as chess, Go, and StarCraft does not necessarily indicate an ability to perform well in open-ended wargames that contain real military problems. See Matthew Walsh, Lance Menthe, Edward Geist, Eric Hastings, Joshua Kerrigan, Jasmin Léveillé, Joshua Margolis, Nicholas Martin, and Brian P. Donnelly, *Exploring the Feasibility and Utility of Machine Learning–Assisted Command and Control: Vol. 1, Findings and Recommendations*, RAND Corporation, RR-A263-1, 2021.

[97] John Curry and Tim Price, *Matrix Games for Modern Wargaming: Developments in Professional and Educational Wargames*, History of Wargaming Project, 2022; John Curry, Chris Engle, and Peter Perla, *The Matrix Games Handbook: Professional Applications from Education to Analysis and Wargaming*, History of Wargaming Project, 2022.

when games are exploratory in nature, the opportunities to develop and employ AI, even if technically feasible, could be prohibitively costly because of the requirements to develop niche technologies to support games with limited repeatability, the limited opportunities for data collection, and the costs associated with resolving improvisational or coconstructed player moves.

Summary

When considering questions of feasibility and cost together, the organizational challenges posed by bringing AI into wargames come into focus. We present our overall conclusion in Figure 4.2. Although many applications may be technically feasible, the costs of bringing them to bear in all cases could be prohibitive. The greatest opportunities for successful applications will remain in the areas of game interpretation, where digital gaming platforms and infrastructure already exist (e.g., alternative conditions and evaluation). Bringing AI into more open-ended and exploratory games will remain costly, particularly given the relative immaturity of AI technologies at handling the requirements for those activities. Note that we do not attempt to estimate the cost of developing human-level AI; we treat that as prohibitive at this point.

Figure 4.2. Feasibility and Cost-Effectiveness of Artificial Intelligence for Wargaming Tasks

Type \ Task	Preparing	Playing	Adjudicating	Interpreting
Systems Exploration	Feasible and affordable when the problem specification is simple and available, especially for repeated games	Feasible for well-defined domains	Feasible when adjudication terms are straightforward	Requires "human-level" AI
Innovation		Requires "human-level" AI	Possibility of unforeseen player actions complicates adjudication	Requires "human-level" AI
Alternative Conditions		Feasible for well-defined domains	Feasible when adjudication terms are straightforward	Feasible except where differences are very complex
Evaluation		Feasible if low-quality players are adequate for game application	Feasible when adjudication terms are straightforward	Feasible except where goals are not well-specified

Prohibitive   Possible   Feasible

To maximize opportunities for success, organizations that have a mandate to regularly develop and employ wargames in routine activities would concentrate on the applications of game-based AI technologies, if only to gain experience and expertise on less difficult applications before moving on to more-difficult ones. Additionally, those organizations would have the resources and demand to invest in technologies that can reduce the overall labor costs associated with developing, performing, and interpreting games. By comparison, employing organizations that conduct open-ended games that are associated with systems exploration and innovation for research purposes to serve as the front line for AI applications could place organizations with more-limited resources and more-difficult technical challenges in a position to fail.

This cost-benefit analysis highlights cases in which AI is most likely and least likely to benefit wargaming. AI is likely to be more useful for alternative conditions or evaluation games than for systems exploration or innovation games. AI could prove particularly useful in games that already give a significant role to computational models during the adjudication process or that generate large volumes of digital information that must be adjudicated (for example, wargames that incorporate cyber or electronic warfare operations).

Games that are played with limited digital infrastructure or interaction with computational models, by contrast, appear to be less promising. AI could prove beneficial for unclassified training seminars, in which advanced HCI can identify patterns of discourse and decisionmaking. In contrast, the exploitation of AI is unlikely to be feasible in classified games that require advanced HCI technologies for data capture and model and asset interaction. AI is much more attractive for the repeated modeling of zero-sum, force-on-force conflicts than for games that are played as one-offs or for a very limited number of times for specific purposes.

Abbreviations

AI	artificial intelligence
DARPA	Defense Advanced Research Projects Agency
HCI	human-computer interaction
ML	machine learning
MCTS	Monte Carlo tree search
RL	reinforcement learning
RSAS	RAND Strategy Assessment System
SME	subject-matter expert
USAF	U. S. Air Force
WOPR	War Operations Plan Response

References

Allen, Thomas B., *War Games: The Secret World of the Creators, Players, and Policy Makers Rehearsing World War III Today*, Berkeley Books, 1989.

Alipourfard, Nazanin, Beatrix Arendt, Daniel Benjamin, Noam Benkler, Mark Burstein, Martin Bush, James Caverlee, Yiling Chen, Chae Clark, Anna Dreber, Timothy M. Errington, Fiona Fidler, Nicholas Fox, Aaron Frank, Hannah Fraser, Scott Friedman, Ben Gelman, James Gentile, C. Lee Giles, Michael Gordon, Reed Gordon-Sarney, Christopher Griffin, Timothy Gulden, Krystal Hahn, Robert Hartman, Felix Holzmeister, Xia Hu, Magnus Johannesson, Lee Kezar, Melissa Kline Struhl, Ugur Kuter, Anthony Kwasnica, Dong-Ho Lee, Kristina Lerman, Yang Liu, Zach Loomas, Bri Luis, Ian Magnusson, Michael Bishop, Olivia Miske, Fallon Mody, Fred Morstatter, Brian A. Nosek, E. Simon Parsons, David Pennock, Thomas Pfeiffer, Haochen Pi, Jay Pujara, Sarah Rajtmajer, Xiang Ren, Abel Salinas, Ravi Selvam, Frank Shipman, Priya Silverstein, Amber Sprenger, Anna Squicciarini, Stephen Stratman, Kexuan Sun, Saatvik Tikoo, Charles R. Twardy, Andrew Tyner, Domenico Viganola, Juntao Wang, David Wilkinson, Bonnie Wintle, and Jian Wu, "Systematizing Confidence in Open Research and Evidence (SCORE)," Center for Open Science, May 2021.

Amazon Web Services, "Amazon Transcribe" website, undated. As of August 27, 2022:
https://aws.amazon.com/transcribe/

Axtell, Robert L., "Short-Term Opportunities, Medium-Run Bottlenecks, and Long-Time Barriers to Progress in the Evolution of an Agent-Based Social Science," in Aaron B. Frank and Elizabeth M. Bartels, eds., *Adaptive Engagement for Undergoverned Spaces: Concepts, Challenges, and Prospects for New Approaches*, RAND Corporation, RR-A1275-1, 2022. As of August 27, 2022:
https://www.rand.org/pubs/research_reports/RRA1275-1.html

Ayoub, Dan, "Unleashing the Power of AI for Education," *MIT Technology Review*, March 4, 2020.

Bartels, Elizabeth M., *Building Better Games for National Security Policy Analysis: Towards a Social Scientific Approach*, dissertation, Pardee RAND Graduate School, RAND Corporation, RGSD-437, 2020. As of August 27, 2022:
https://www.rand.org/pubs/rgs_dissertations/RGSD437.html

Bartels, Elizabeth M., Aaron B. Frank, Jasmin Léveillé, Timothy Marler, and Yuna Huh Wong, "Gaming Undergoverned Spaces: Emerging Approaches for Complex National Security Policy Problems," in Aaron B. Frank and Elizabeth M. Bartels, eds., *Adaptive Engagement for Undergoverned Spaces: Concepts, Challenges, and Prospects for New Approaches*, RAND Corporation, RR-A1275-1, 2022. As of August 27, 2022:
https://www.rand.org/pubs/research_reports/RRA1275-1.html

Bartels, Elizabeth M., Igor Mikolic-Torreira, Steven W. Popper, and Joel B. Predd, *Do Differing Analyses Change the Decision? Using a Game to Assess Whether Differing Analytic Approaches Improve Decisionmaking*, RAND Corporation, RR-2735-RC, 2019. As of August 25, 2022:
https://www.rand.org/pubs/research_reports/RR2735.html

Bergdahl, Jacob, "An AI Wrote This Story," *Medium*, July 16, 2021.

Berkeley, Edmund C., *Giant Brains; or, Machines That Think*, John Wiley & Sons, 1949.

Brewer, Garry D., and Martin Shubik, *The War Game: A Critique of Military Problem Solving*, Harvard University Press, 1979.

Brodie, Bernard, *The American Scientific Strategists*, RAND Corporation, P-2979, 1964. As of August 27, 2022:
https://www.rand.org/pubs/papers/P2979.html

Brown, Thomas A., and Edwin W. Paxson, *A Retrospective Look at Some Strategy and Force Evaluation Games*, RAND Corporation, R-1619-PR, 1975. As of August 26, 2022:
https://www.rand.org/pubs/reports/R1619.html

Cassel, David, "Can We Teach an AI to Play Dungeons and Dragons?" *The New Stack*, March 28, 2021.

Chollet, François, "On the Measure of Intelligence," arXiv, November 5, 2019.

Connable, Ben, Michael J. McNerney, William Marcellino, Aaron B. Frank, Henry Hargrove, Marek N. Posard, S. Rebecca Zimmerman, Natasha Lander, Jasen J. Castillo, and James Sladden, *Will to Fight: Analyzing, Modeling, and Simulating the Will to Fight of Military Units*, RAND Corporation, RR-2341-A, 2018. As of August 27, 2022:
https://www.rand.org/pubs/research_reports/RR2341.html

Crevier, Daniel, *AI: The Tumultuous History of the Search for Artificial Intelligence*, Basic Books, 1993.

Curry, John, Chris Engle, and Peter Perla, *The Matrix Games Handbook: Professional Applications from Education to Analysis and Wargaming*, History of Wargaming Project, 2022.

Curry, John, and Tim Price, *Matrix Games for Modern Wargaming: Developments in Professional and Educational Wargames*, History of Wargaming Project, 2022.

DARPA—*See* Defense Advanced Research Projects Agency.

Davis, Paul K., *RAND's Experience in Applying Artificial Intelligence Techniques to Strategic-Level Military-Political War Gaming*, RAND Corporation, P-6977, 1984. As of August 27, 2022:
https://www.rand.org/pubs/papers/P6977.html

Davis, Paul K., "Knowledge-Based Simulation for Studying Issues of Nuclear Strategy," in Allan M. Din, ed., *Arms and Artificial Intelligence: Weapon and Arms Control Applications of Advanced Computing*, Oxford University Press, 1987.

Davis, Paul K., "Some Lessons Learned from Building Red Agents in the RAND Strategy Assessment System (RSAS)," RAND Corporation, N-3003-OSD, 1989. As of December 13, 2022:
https://www.rand.org/pubs/notes/N3003.html

Davis, Paul K., "Synthetic Cognitive Modeling of Adversaries for Effects-Based Planning," *Enabling Technologies for Simulation Science VI*, Vol. 4716, July 2002.

Davis, Paul K., "Illustrating a Model-Game-Model Paradigm for Using Human Wargames in Analysis," RAND Corporation, WR-1179, 2017. As of August 27, 2022:
https://www.rand.org/pubs/working_papers/WR1179.html

Davis, Paul K., and Paul Bracken, "Artificial Intelligence for Wargaming and Modeling," *Journal of Defense Modeling and Simulation*, February 2022.

Davis, Paul K., Angela O'Mahony, Christian Curriden, and Jonathan Lamb, *Influencing Adversary States: Quelling Perfect Storms*, RAND Corporation, RR-A161-1, 2021. As of February 24, 2023:
https://www.rand.org/pubs/research_reports/RRA161-1.html

Defense Advanced Research Projects Agency, "Assault Breaker," webpage, undated-a. As of October 24, 2022:
https://www.darpa.mil/about-us/timeline/assault-breaker

Defense Advanced Research Projects Agency, "Explainable Artificial Intelligence (XAI) (Archived)," webpage, undated-b. As of August 30, 2022:
https://www.darpa.mil/program/explainable-artificial-intelligence

Din, Allan M., ed., *Arms and Artificial Intelligence: Weapon and Arms Control Applications of Advanced Computing*, Oxford University Press, 1987.

Dreyfus, Hubert L., *Alchemy and Artificial Intelligence*, RAND Corporation, P-3244, 1965. As of August 25, 2022:
https://www.rand.org/pubs/papers/P3244.html

Elias, George Skaff, Richard Garfield, and K. Robert Gutschera, *Characteristics of Games*, MIT Press, 2012.

Fokker, Jeroen, "The Chess Example in Turing's Mind Paper Is Really About Ambiguity," in Andrei Voronkov, ed., *Turing-100: The Alan Turing Centenary*, EPiC Series in Computing, Vol. 10, June 2012.

Frank, Aaron B., and Elizabeth M. Bartels, eds., *Adaptive Engagement for Undergoverned Spaces: Concepts, Challenges, and Prospects for New Approaches*, RAND Corporation, RR-A1275-1, 2022. As of October 3, 2023:
https://www.rand.org/pubs/research_reports/RRA1275-1.html

Gehlhaus, Diana, Luke Koslosky, Kayla Goode, and Claire Perkins, *U.S. AI Workforce: Policy Recommendations*, Center for Security and Emerging Technology, October 2021.

Goodman, James, Sebastian Risi, and Simon Lucas, "AI and Wargaming," arXiv, September 25, 2020.

Google Cloud, "Explainable AI," website, undated. As of August 27, 2022:
https://cloud.google.com/explainable-ai

GPT-3, "A Robot Wrote This Entire Article. Are You Scared Yet, Human?" *The Guardian*, September 8, 2020.

Grana, Justin, "Difficulties in Analyzing Strategic Interaction: Quantifying Complexity," in Aaron B. Frank and Elizabeth M. Bartels, eds., *Adaptive Engagement for Undergoverned Spaces: Concepts, Challenges, and Prospects for New Approaches*, RAND Corporation, RR-A1275-1, 2022. As of August 27, 2022:
https://www.rand.org/pubs/research_reports/RRA1275-1.html

Graubard, Morlie Hammer, and Carl H. Builder, *New Methods for Strategic Analysis: Automating the Wargame*, RAND Corporation, P-6763, 1982. As of August 26, 2022:
https://www.rand.org/pubs/papers/P6763.html

Gulden, Timothy R., Jonathan Lamb, Jeff Hagen, and Nicholas A. O'Donoughue, *Modeling Rapidly Composable, Heterogeneous, and Fractionated Forces: Findings on Mosaic Warfare from an Agent-Based Model*, RAND Corporation, RR-4396-OSD, 2021. As of August 27, 2022:
https://www.rand.org/pubs/research_reports/RR4396.html

Higham, Nicholas J., ed., *The Princeton Companion to Applied Mathematics*, Princeton University Press, 2016.

Hintze, Arend, "Open-Endedness for the Sake of Open-Endedness," *Artificial Life*, Vol. 25, No. 2, Spring 2019.

Icosystem, "Hunch Engine," webpage, undated. As of October 4, 2023: https://web.archive.org/web/20230208195425/http://www.icosystem.com:80/technology/hunch-engine/

Kelly, George, and Hugh McCabe, "A Survey of Procedural Techniques for City Generation," *ITB Journal*, Vol. 7, No. 2, 2006.

Knox, MacGregor, and Williamson Murray, eds., *The Dynamics of Military Revolution, 1300–2050*, Cambridge University Press, 2001.

Langley, Patrick W., Herbert A. Simon, Gary Bradshaw, and Jan M. Zytkow, *Scientific Discovery: Computational Explorations of the Creative Process*, MIT Press, 1987.

Launchbury, John A., "DARPA Perspective on Artificial Intelligence," briefing slides, Defense Advanced Research Projects Agency, undated.

Lingel, Sherrill, Jeff Hagen, Eric Hastings, Mary Lee, Matthew Sargent, Matthew Walsh, Li Ang Zhang, and David Blancett, *Joint All-Domain Command and Control for Modern Warfare: An Analytic Framework for Identifying and Developing Artificial Intelligence Applications*, RAND Corporation, RR-4408/1-AF, 2020. As of August 27, 2022: https://www.rand.org/pubs/research_reports/RR4408z1.html

Lohn, Andrew J., and Micah Musser, *AI and Compute: How Much Longer Can Computing Power Drive Artificial Intelligence Progress?* Center for Security and Emerging Technology, January 2022.

Marshall, Andrew W., J. J. Martin, and Henry S. Rowen, eds., *On Not Confusing Ourselves: Essays on National Security Strategy in Honor of Albert and Roberta Wohlstetter*, Westview Press, 1991.

Marto, Anabela, and Alexandrino Gonçalves, "Augmented Reality Games and Presence: A Systematic Review," *Journal of Imaging*, Vol. 8, No. 4, April 2022.

Menthe, Lance, Dahlia Anne Goldfeld, Abbie Tingstad, Sherrill Lingel, Edward Geist, Donald Brunk, Amanda Wicker, Sarah Lovell, Balys Gintautas, Anne Stickells, and Amado Cordova, *Technology Innovation and the Future of Air Force Intelligence Analysis*: Vol. 2, *Technical Analysis and Supporting Material*, RAND Corporation, RR-A341-2, 2021. As of October 4, 2023: https://www.rand.org/pubs/research_reports/RRA341-2.html

Menthe, Lance, Li Ang Zhang, Edward Geist, Joshua Steier, Aaron B. Frank, Eric Van Hegewald, Gary J. Briggs, Keller Scholl, Yusuf Ashpari, and Anthony Jacques, *Understanding the Limits of Artificial Intelligence for Warfighters*: Vol. 1, *Summary*, RR-A1722-1, 2024.

Microsoft, "Holoportation," website, undated. As of August 27, 2022: https://www.microsoft.com/en-us/research/project/holoportation-3/

Millot, Marc Dean, Roger C. Molander, and Peter A. Wilson, *"The Day After . . ." Study: Nuclear Proliferation in the Post-Cold War World-Volume II, Main Report*, RAND Corporation, MR-253-AF, 1993. As of August 27, 2022: https://www.rand.org/pubs/monograph_reports/MR253.html

Minsky, Marvin, ed., *Semantic Information Processing*, MIT Press, 1968.

Murray, Williamson, and MacGregor Knox, "Conclusion: The Future Behind Us," in MacGregor Knox and Williamson Murray, eds., *The Dynamics of Military Revolution, 1300–2050*, Cambridge University Press, 2001.

Murray, Williamson, and Allan R. Millett, eds., *Military Innovation in the Interwar Period*, Cambridge University Press, 1998.

Nilsson, Nils J., *The Quest for Artificial Intelligence*, Cambridge University Press, 2009.

Osborne, Martin J., *An Introduction to Game Theory*, Oxford University Press, 2004.

Ouimet, Kirk, "Conversations with GPT-3," Medium, July 16, 2020.

Paul, Debjit, and Anette Frank, "COINS: Dynamically Generating COntextualized Inference Rules for Narrative Story Completion," arXiv, June 4, 2021.

Perla, Peter P., Michael Markowitz, Albert Nofi, Christopher Weuve, Julia Loughran, and Marcy Stahl, *Gaming and Shared Situation Awareness*, Center for Naval Analyses, November 2000.

Plaat, Aske, Walter Kosters, and Jaap van den Herik, eds., *Computers and Games: 9th International Conference, CG 2016, Leiden, The Netherlands, June 29–July 1, 2016, Revised Selected Papers*, Vol. 10068, Springer, 2016.

Ramesh, Aditya, Mikhail Pavlov, Gabriel Goh, Scott Gray, Chelsea Voss, Alec Radford, Mark Chen, and Ilya Sutskever, "Zero-Shot Text-to-Image Generation," *Proceedings of the 38th International Conference on Machine Learning*, Proceedings of Machine Learning Research, Vol. 139, 2021.

Reed, Scott, Konrad Żołna, Emilio Parisotto, Sergio Gómez Colmenarejo, Alexander Novikov, Gabriel Barth-Maron, Mai Giménez, Yury Sulsky, Jackie Kay, Jost Tobias Springenberg, Tom Eccles, Jake Bruce, Ali Razavi, Ashley Edwards, Nicolas Heess, Yutian Chen, Raia Hadsell, Oriol Vinyals, Mahyar Bordbar, and Nando de Freitas, "A Generalist Agent," *Transactions on Machine Learning Research*, November 2022.

Rieber, Steven, "Methods for Evaluating Analytic Quality of Reasoning RFI Number IARPA-RFI-22-02," Intelligence Advanced Research Projects Activity, November 30, 2021.

Rodriguez, Jesus, "Introducing AI Explainability 360: A New Toolkit to Help You Understand What Machine Learning Models Are Doing," KDnuggets, August 27, 2019.

Rosen, Stephen Peter, "Net Assessment as an Analytical Concept," in Andrew W. Marshall, J. J. Martin, and Henry S. Rowen, eds., *On Not Confusing Ourselves: Essays on National Security Strategy in Honor of Albert and Roberta Wohlstetter*, Westview Press, 1991.

Rosen, Stephen Peter, *Winning the Next War: Innovation and the Modern Military*, Cornell University Press, 1994.

Russell, Laurence, "Mirrorscape Wants to Conjure Your Favorite Tabletop Game in AR," *Wired*, January 18, 2022.

Samuel, Arthur L., "Some Studies in Machine Learning Using the Game of Checkers," *IBM Journal of Research and Development*, Vol. 3, No. 3, July 1959.

Sawyer, R. Keith, *Social Emergence: Societies as Complex Systems*, Cambridge University Press, 2005.

Schaeffer, Jonathan, Neil Burch, Yngvi Björnsson, Akihiro Kishimoto, Martin Müller, Robert Lake, Paul Lu, and Steve Sutphen, "Checkers Is Solved," *Science*, Vol. 317, No. 5844, September 14, 2017.

Scholl, Keller, Gary Briggs, Li Ang Zhang, and John Salmon, *Understanding the Limits of Artificial Intelligence for Warfighters:* Vol. 5, *Mission Planning,* RR-A1722-5, 2024.

Schuemie, Martijn J., Peter van der Straaten, Merel Krijn, and Charles A. P. G. van der Mast, "Research on Presence in Virtual Reality: A Survey," *CyberPsychology & Behavior,* Vol. 4, No. 2, July 2004.

Shannon, Claude E., "XXII. Programming a Computer for Playing Chess," *London, Edinburgh, and Dublin Philosophical Magazine and Journal of Science,* Vol. 41, No. 314, 1950.

Silver, David, Thomas Hubert, Julian Schrittwieser, Ioannis Antonoglou, Matthew Lai, Arthur Guez, Marc Lanctot, Laurent Sifre, Dharshan Kumaran, Thore Graepel, Timothy Lillicrap, Karen Simonyan, and Demis Hassabis, "A General Reinforcement Learning Algorithm That Masters Chess, Shogi, and Go Through Self-Play," *Science,* Vol. 362, No. 6419, December 7, 2018.

Simon, Herbert A., and Allen Newell, "Heuristic Problem Solving: The Next Advance in Operations Research," *Operations Research,* Vol. 6, No. 1, January–February1958.

Simtable, "About Simtable," webpage, undated. As of June 12, 2020: http://www.simtable.com/about/

Singler, Beth, "Dungeons and Dragons, Not Chess and Go: Why AI Needs Roleplay," Aeon, April 3, 2018.

Stanley, Kenneth O., Joel Lehman, and Lisa Soros, "Open-Endedness: The Last Grand Challenge You've Never Heard of," O'Reilly Media, December 19, 2017.

Steier, Joshua, Erik Van Hegewald, Anthony Jacques, Gavin S. Hartnett, and Lance Menthe, *Understanding the Limits of Artificial Intelligence for Warfighters:* Vol. 2, *Distributional Shift in Cybersecurity Datasets,* RR-A1722-2, 2024.

Stokel-Walker, Chris, and Richard Van Noorden, "What ChatGPT and Generative AI Mean for Science," *Nature,* February 6, 2023.

Sutton, Richard S., and Andrew G. Barto, *Reinforcement Learning: An Introduction,* 1st ed., MIT Press, 1998.

Tacla, Joaquin Victor, "Air Force Pilots Will Use AR Headsets to Fight AI-Powered Enemies!" Tech Times, August 5, 2022.

Verma, Pranshu, "Fighter Pilots Will Don Augmented Reality Helmets for Training," *Washington Post,* August 4, 2022.

Vinyals, Oriol, Igor Babuschkin, Wojciech M. Czarnecki, Michaël Mathieu, Andrew Dudzik, Junyoung Chung, David H. Choi, Richard Powell, Timo Ewalds, Petko Georgiev, Junhyuk Oh, Dan Horgan, Manuel Kroiss, Ivo Danihelka, Aja Huang, Laurent Sifre, Trevor Cai, John P. Agapiou, Max Jaderberg, Alexander S. Vezhnevets, Rémi Leblond, Tobias Pohlen, Valentin Dalibard, David Budden, Yury Sulsky, James Molloy, Tom L. Paine, Caglar Gulcehre, Ziyu Wang, Tobias Pfaff, Yuhuai Wu, Roman Ring, Dani Yogatama, Dario Wünsch, Katrina McKinney, Oliver Smith, Tom Schaul, Timothy Lillicrap, Koray Kavukcuoglu, Demis Hassabis, Chris Apps, and David Silver, "Grandmaster Level in StarCraft II Using Multi-Agent Reinforcement Learning," *Nature,* Vol. 575, November 14, 2019.

Voronkov, Andrei, ed., *Turing-100: The Alan Turing Centenary,* EPiC Series in Computing, Vol. 10, June 2012.

Wabiński, Jakub, and Albina Mościcka, "Automatic (Tactile) Map Generation—A Systematic Literature Review," *ISPRS International Journal of Geo-Information,* Vol. 8, No. 7, 2019.

Walsh, Matthew, Lance Menthe, Edward Geist, Eric Hastings, Joshua Kerrigan, Jasmin Léveillé, Joshua Margolis, Nicholas Martin, and Brian P. Donnelly, *Exploring the Feasibility and Utility of Machine Learning-Assisted Command and Control: Vol. 1, Findings and Recommendations*, RAND Corporation, RR-A263-1, 2021. As of February 24, 2023:
https://www.rand.org/pubs/research_reports/RRA263-1.html

Wang, Elaine Lin, Lindsay Clare Matsumura, Diane Litman, Richard Correnti, Haoran Zhang, Zahra Rahimi, Zahid Kisa, Ahmed Magooda, Emily Howe, and Rafael Quintana, *Contributions to Research on Automated Writing Scoring and Feedback Systems*, RAND Corporation, RB-A1062-1, 2022. As of August 27, 2022:
https://www.rand.org/pubs/research_briefs/RBA1062-1.html

WarGames, dir. John Badham, United Artists, 1983.

Wasser, Becca, Jenny Oberholtzer, Stacie L. Pettyjohn, and William Mackenzie, *Gaming Gray Zone Tactics: Design Considerations for a Structured Strategic Game*, RAND Corporation, RR-2915-A, 2019. As of August 27, 2022:
https://www.rand.org/pubs/research_reports/RR2915.html

Watts, Barry, and Williamson Murray, "Military Innovation in Peacetime," in Williamson Murray and Allan R. Millett, eds., *Military Innovation in the Interwar Period*, Cambridge University Press, 1998.

Weiner, Milton G., *War Gaming Methodology*, RAND Corporation, RM-2413, 1959. As of August 26, 2022:
https://www.rand.org/pubs/research_memoranda/RM2413.html

Wen, Mingyun, Jisun Park, and Kyungeun Cho, "A Scenario Generation Pipeline for Autonomous Vehicle Simulators," *Human-centric Computing and Information Sciences*, Vol. 10, 2020.

White, Richard H., Edward Smith, An-Jen Tai, William E. Cralley, Joel Christenson, David Davis, William Fedorochko, Jr., Jack LeCuyer, Dayton Maxwell, Klaus Niemeyer, Edwin Pechous, Danielle Philips, Ian Rehmert, Marc Samuels, and Katherine Ward, *An Introduction to IDA's S.E.N.S.E.—R.S.A. Project*, Institute for Defense Analyses, September 1999.

Wiener, Norbert, *The Human Use of Human Beings: Cybernetics and Society*, Houghton Mifflin, 1950.

Wolfe, Bernard, "Self Portrait," *Galaxy Science Fiction*, Vol. 3, No. 2, November 1951.

Wolfe, Bernard, *Limbo*, Hachette, 1952.

Zhang, Li Ang, Yusuf Ashpari, and Anthony Jacques, *Understanding the Limits of Artificial Intelligence for Warfighters: Vol. 3, Predictive Maintenance*, RR-A1722-3, 2024.